New Insights into Solar Energy

New Insights into Solar Energy

Editor: Catherine Waltz

STATES
ACADEMIC PRESS
www.statesacademicpress.com

States Academic Press,
109 South 5th Street,
Brooklyn, NY 11249, USA

Visit us on the World Wide Web at:
www.statesacademicpress.com

ISBN: 978-1-63989-380-5 (Hardback)

Cataloging-in-Publication Data

New insights into solar energy / edited by Catherine Waltz.
　　p. cm.
Includes bibliographical references and index.
ISBN 978-1-63989-380-5
1. Solar energy. 2. Renewable energy sources. I. Waltz, Catherine.
TJ810 .N49 2022
621.47--dc23

Table of Contents

Preface

This book was inspired by the evolution of our times; to answer the curiosity of inquisitive minds. Many developments have occurred across the globe in the recent past which has transformed the progress in the field.

Solar energy is the power obtained by the conversion of heat and light radiated by sun using techniques such as photovoltaics, solar heating and artificial photosynthesis. It is a renewable source of energy which is also believed to be the cleanest and most abundant renewable energy source. Solar technologies are classified into active and passive solar systems. Active solar systems comprise devices to convert sun's energy into a more usable form such as electricity or hot water. Passive solar techniques do not employ any external electrical or mechanical device and focus on optimizing the use of heat or light directly from the sun. The energy captured using solar devices can be employed for various purposes such as lighting, cooking, heating, transportation, etc. This book traces the progress of this field and highlights some of its key concepts and applications. It will also provide interesting topics for research which interested readers can take up. As this field is emerging at a rapid pace, the contents of this book will help the readers understand the modern concepts and applications of solar energy.

This book was developed from a mere concept to drafts to chapters and finally compiled together as a complete text to benefit the readers across all nations. To ensure the quality of the content we instilled two significant steps in our procedure. The first was to appoint an editorial team that would verify the data and statistics provided in the book and also select the most appropriate and valuable contributions from the plentiful contributions we received from authors worldwide. The next step was to appoint an expert of the topic as the Editor-in-Chief, who would head the project and finally make the necessary amendments and modifications to make the text reader-friendly. I was then commissioned to examine all the material to present the topics in the most comprehensible and productive format.

I would like to take this opportunity to thank all the contributing authors who were supportive enough to contribute their time and knowledge to this project. I also wish to convey my regards to my family who have been extremely supportive during the entire project.

Editor

Floating Solar Chimney Technology

Christos D. Papageorgiou
National Technical University of Athens
Greece

1. Introduction

1.1 Floating Solar Chimney technology description

The purpose of this chapter is to present the Floating solar chimney (FSC) technology, look for the site www.floatingsolarchimney.gr, in order to explain its principles of operation and to point out its various significant benefits. This technology is the advisable one for candidacy for large scale solar electricity generation especially in desert or semi desert areas of our planet and a major technology for the global warming elimination.

The solar chimney power plants are usually referred to as solar updraft towers (http://en.wikipedia.org/wiki/Solar_updraft_tower) and the related solar chimneys are huge reinforced concrete structures. However due to the high construction cost of the concrete solar chimneys the solar up-draft tower technology is expensive demanding a high initial investment in comparison to its competitive solar technologies. Their solar up-draft towers are huge structures of high initial investment cost that can not be split into small units. That is possible for the relatively also expensive PV solar technology. Also the solar updraft technology is far more expensive compared to the conventional fossil fueled power plants of similar electricity generation. That is why the solar chimney technology has not yet been applied although it is a solar technology of many advantages.

The **Floating Solar Chimney (FSC)** is a fabric low cost alternative of the concrete solar chimney up-draft towers that can make the Floating Solar Chimney technology cost competitive in comparison not only with the renewable electricity generation technologies but also with the conventional fossil fueled electricity generation technologies. Also the FSC technology is cost effective to be split into small units of several MW each.

The Floating Solar Chimney Power Plant, named by the author as **Solar Aero-Electric Power Plant (SAEP)** due to its similarity to the Hydro-Electric power plant, is a set of three major components:

- **The Solar Collector.** It is a large greenhouse open around its periphery with a transparent roof supported a few meters above the ground.
- **The Floating Solar Chimney (FSC).** It is a tall fabric cylinder placed at the centre of the solar collector through which the warm air of the greenhouse, due to its relative buoyancy to the ambient air, is up-drafting.
- **The Turbo-Generators.** It is a set of air turbines geared to appropriate electric generators in the path of up-drafting warm air flow that are forced to rotate generating electricity. The gear boxes are adjusting the rotation speed of the air turbines to the generator rotation speed defined by the grid frequency and their pole pairs.

An indicative figure of a solar chimney Power Plant with a circular solar collector and a Floating Solar Chimney inclined due to external winds is shown in next figure(1).

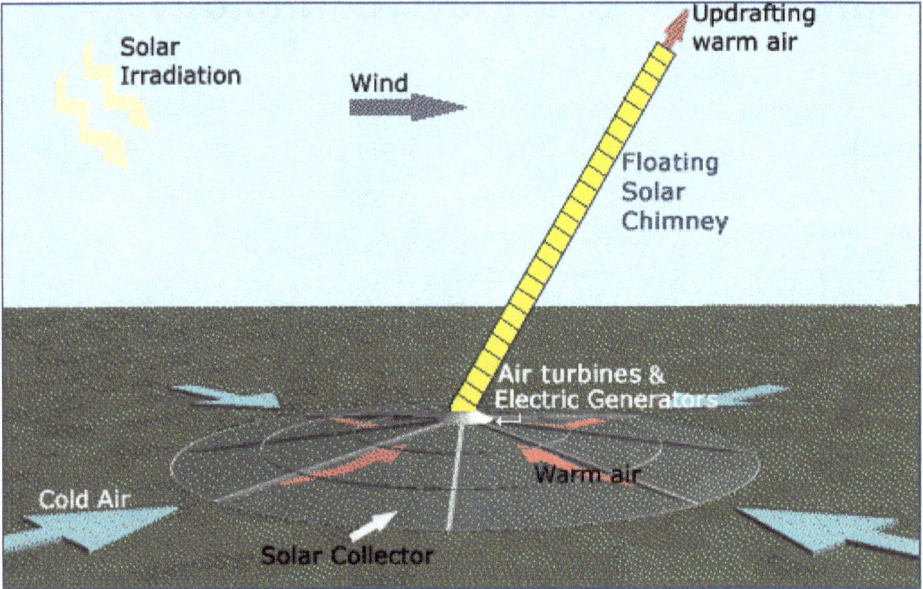

Fig. 1. Floating Solar Chimney Power Plant in operation

Because of its patented construction the FSC is a free standing lighter than air structure that can tilt when external winds appear. Low cost Floating Solar Chimneys up to 1000 m with internal diameters 25 m ÷ 40 m, can be constructed with an existing polyester fabric, giving to their respective Solar Aero-Electric Power Plants, low investment costs.

By this innovating Floating Solar Chimney Technology of heights of the FSCs up to1000m, up to 1.2 % of the arriving horizontal solar radiation on the solar collector surface, can be converted to electricity

1.2 Similarity to hydro-electric power plants

The Floating Solar Chimney power plants, due to their similarity to hydro-electric power plants, are named by the author Solar Aero Electric Power Plants (SAEPs).

Their similarity is due to the following facts:

- The hydro-electric PPs operate due to falling water gravity, while the solar aero-electric PPs operate due to the up-drafting warm air buoyancy.
- The electricity generation units of hydro-electric PPs are water turbines engaged to electric generators while the generation units of solar aero-electric PPs are air turbines engaged to electric generators.
- The energy produced by the hydro-electric PPs is proportional to the falling water height, while the energy produced by the solar aero-electric PPs is proportional to up-drafting height of warm air, which is equal to the height of the solar chimneys.
- That is why Prof J. Sclaigh in his book named the solar chimney technology power plants as the hydro-electric power plants of deserts.

1.3 Continuous operation

As it will be shown later the SAEPs operate continuously due to the ground thermal storage. The minimum electric power is generated when the sun is just starting rising, while the maximum electric power is achieved about 2 hours after the sun's maximum irradiation on ground. The power generation profile can become smoother if we increase the solar collector thermal capacity. This can be done by putting on its ground area closed tubes filled with water (as happens already in conventional greenhouses).

2. History

The Solar Chimney technology for electricity generation was inspired by several engineering pioneers early in the first decade of the 20th century.

In 1926 Prof Engineer Bernard Dubos proposed to the French Academy of Sciences the construction of a Solar Aero-Electric Power Plant in North Africa with its solar chimney on the slope of a sufficient height mountain. His proposal is shown in the following figure(2), found in a book of 1954 ("Engineer's Dream" Willy Ley, Viking Press 1954)

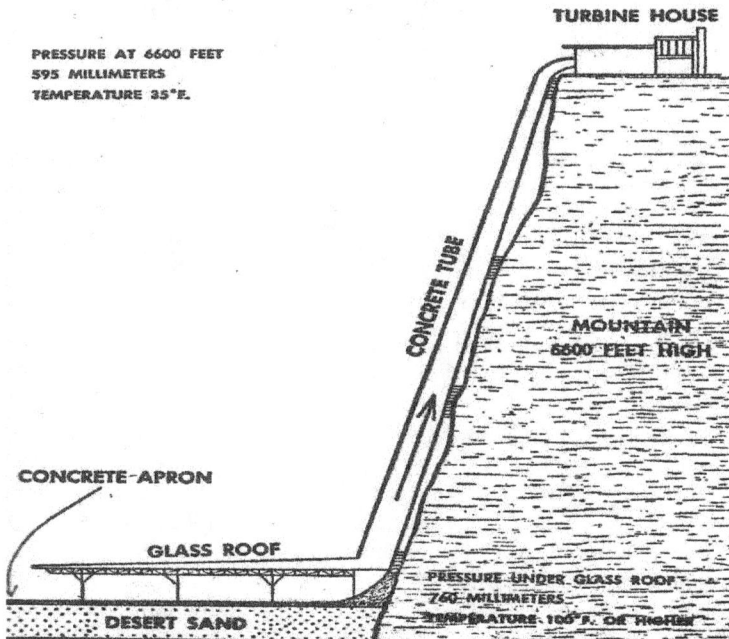

Fig. 2. (from the book: "Engineer's Dream"By: Willy Ley, Viking Press 1954)

Lately Schaich, Bergerman and Partners, under the direction of Prof. Dr. Ing. Jorg Schlaigh, built an operating model of a SAEPP in 1982 in Manzaranes (Spain), which was funded by the German Government.

This solar chimney power plant, shown in next figure (3) was of rating power 50 KW. Its greenhouse had a surface area of 46000 m^2 and its solar chimney was made out of steel tubes of 10 m diameter and had a height of 195 m.

This demo SAEP was operating successfully for approximately 6 years. During its operation, optimization data were taken.

The collected operational data were in accordance with the theoretical results of the scientific team of Prof Jorg Schlaigh.

Fig. 3. A view of the Manzanares Solar Chimney Power Plant

Prof. Jorg Schlaigh in 1996 published a book (Schlaigh 1995) presenting the solar chimney technology. He proposed in his book the huge reinforced concrete solar chimneys of heights of 500m-1000m.

The proposed concrete solar chimneys are huge and very expensive. Therefore the investment cost per produced KWh on the solar chimney technology with concrete chimneys is in the same cost range with the competitive solar thermal technologies. The generated KWh, by the CSP Parabolic Through for example, it has almost the same direct production cost, but the CSP power plants can be split into small units and developed using reasonable recourses.

However the proposed solar chimney technology had an important benefit in comparison with the major renewable technologies (Wind, SCP, PV).

That is its ability, equipping its solar collectors, with thermal storage facilities of negligible cost, to generate uninterrupted electricity of a controlled smooth profile for 24h/day, 365days/year.

The last decade several business plans and a series of scientific research papers have focused on the solar chimney technology, whereby the author with a series of patents and papers has introduced and scientifically supported the floating solar technology (Papageorgiou 2004, 2009).

3. Principles of operation of the solar chimney technology and its annual efficiency Information

3.1 Short description and principles of operation

A floating solar chimney power plant (SAEP) is made of three major components:

- A large solar collector, usually circular, which is made of a transparent roof supported a few meters above the ground (the greenhouse). The transparent roof can be made of glass or crystal clear plastic. A second cover made of thin crystal clear plastic is suggested to be hanged just underneath the roof in order to increase its thermal efficiency. The periphery of the solar collector is open in order that the ambient air can move freely into it.

- A tall fabric free standing lighter than air cylinder (the floating solar chimney) placed in the center of the greenhouse which is up drafting the warm air of the greenhouse, due to its buoyancy, to the upper atmospheric layers.

- A set of air turbines geared to appropriate electric generators (the turbo generators), placed with a horizontal axis in a circular path around the base of the FSC or with a vertical axis inside the entrance of the solar chimney. The air turbines are caged and can be just a rotor with several blades or a two stage machine (i.e. with a set of inlet guiding vanes and a rotor of several blades). The gear boxes are adjusting the rotation frequency of the air turbines to the electric generator rotation frequency defined by the grid frequency and the electric generator pole pairs.

The horizontal solar irradiation passing through the transparent roof of the solar collector is heating the ground beneath it. The air beneath the solar collector is becoming warm through a heat transfer process from the ground area to the air. This heat transfer is increased due to the greenhouse effect of the transparent roof.

This warm air becomes lighter than the ambient air. The buoyancy of the warm air is forcing the warm air to escape through the solar chimney. As the warm air is up drafting through the chimney, fresh ambient air is entering from the open periphery of the greenhouse. This fresh air becomes gradually warm, while moving towards the bottom of the solar chimney, and it is also up-drafting.

Thus a large quantity of air mass is continuously circulating from ground to the upper layers of the atmosphere. This circulating air mass flow is offering a part of its thermodynamic energy to the air turbines which rotate and force the geared electric generators also to rotate. Thus the rotational mechanical power of the air turbines is transformed to electrical power. An indicative diagram of the SAEP operation is shown in the next figure(4).

Thus the first two parts of the SAEPs form a huge thermodynamic device up drafting the ground ambient air to the upper atmosphere layers and the third part of the SAEP is the electricity generating unit.

The solar energy arriving on the horizontal surface area A_c of the greenhouse of the SAEP is given by $E_{IR}=A_c \cdot W_y$, where W_y is the annual horizontal solar irradiation in KWh/m^2, at the place of installation of the SAEP and is given by the meteorological data nearly everywhere. The average annual horizontal solar irradiance is given by $G_{av}=W_y/A_c$.

The horizontal solar irradiation is offering thermal power $P_{Th}= \dot{m} \cdot c_p \cdot (T_{03}-T_{02})$ to the up drafting air mass flow $\dot{m}$ of the ambient air, $c_p \approx 1005$ and T_{02} is equal to the average ambient temperature T_0 plus $\sim 0.5 \ ^0K$, in order that it is taken into account the outer air stream increased inlet temperature due to its proximity to the ground on its entrance inside the solar collector.

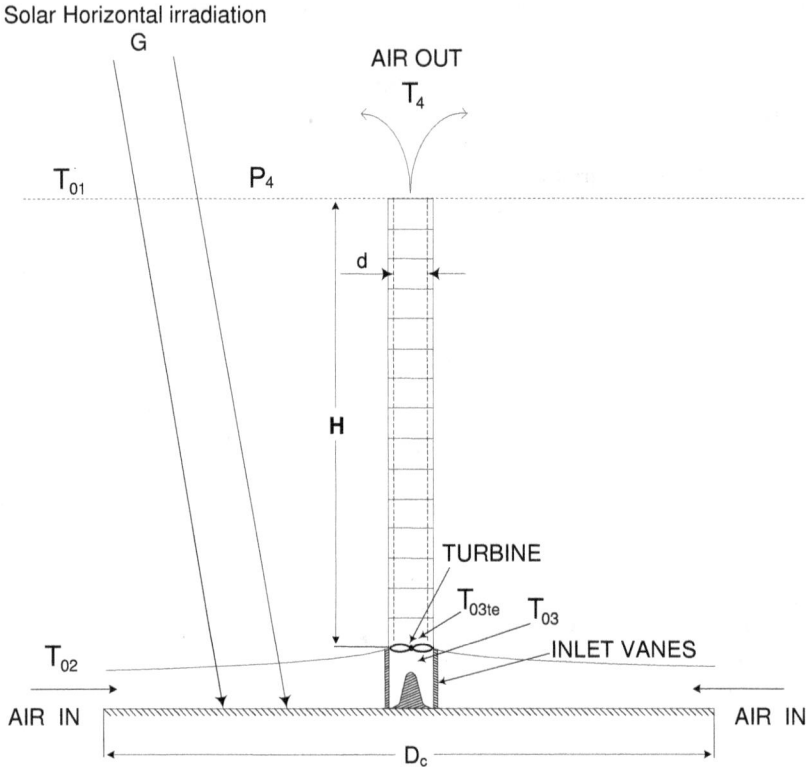

Fig. 4. Schematic diagram of the SAEP in operation

3.2 Annual average efficiency of SAEPs

The annual efficiency of the solar collector η_{sc} is defined as the average ratio of the thermal power P_{Th} absorbed by the air mass flow to the horizontal solar irradiation arriving on the greenhouse roof $G_{av} \cdot A_c$, where G_{av} is the average horizontal irradiance and A_c the greenhouse surface area.

The annual average double glazing solar collector efficiency η_{sc} is theoretically estimated to ~50%, while the annual efficiency for the single glazing solar collector is estimated to 2/3 of the previous figure i.e. ~33%.

Thus the average exit temperature T_{03} from the solar collector can be calculated by the equation $\dot{m} \cdot c_p \cdot (T_{03} - T_{02}) = \eta_{sc} \cdot G_{av} \cdot A_c$ where T_{02} is the average inlet air temperature.

The exit thermal power P_{Th} from the solar collector is transformed to electric power P, plus power thermal losses P_L (to the air turbines, gear boxes and electric generators), plus warm air kinetic power at the top exit of the solar chimney P_{KIN} and friction thermal losses inside the solar chimney P_{FR}.

The maximum efficiency of the solar chimney is the Carnot efficiency defined as the ratio of the temperature difference between the incoming and outcoming air temperatures of the up-drafting air divided by the ambient air temperature.

This maximum efficiency has been proven (Gannon & Backstrom 2000) to be equal to:

$$\eta_{FSC,max}=g \cdot H/(c_p \cdot T_0) \qquad (1)$$

Due to friction and kinetic losses in the solar chimney the actual solar chimney efficiency η_{FSC} is for a properly designed SAEP approximately 90% of its maximum Carnot efficiency (close to the optimum point of operation of the SAEP).

The combined efficiency η_T of the air turbines, gear boxes and electric generators is within the range of 80%.

The average annual efficiency of the SAEP is the product of the average efficiencies of its three major components i.e. the solar collector, the floating solar chimney and the turbo-generators i.e. $\eta_{av}= \eta_{sc} \cdot \eta_{FSC} \cdot \eta_T$.

Thus the annual average efficiency of a SAEP of proper design, with a double glazing solar collector should be approximately:

$$\eta_{av} =(1.2 \cdot H/1000) \% \qquad (2)$$

While for the SAEP with a single cover collector it is approximately:

$$\eta_{av}= (0.79 \cdot H/1000) \%. \qquad (3)$$

The formulae have been calculated for $g=9.81, c_p=1005$ and $T_0 \approx 293.2^0K(20^0C)$.

This means that if the annual horizontal irradiation arriving on the place of installation of the SAEP is 2000 KWh/m^2, the solar collector surface area is 10^6 m^2 (one square Km) and the solar chimney height is 750 m the SAEP can generate approximately 18 million KWh. The same SAEP with a single glazing roof will generate approximately only 12 million KWh.

Following approximate analysis, for a SAEP with a double cover roof of given dimensions (Ac=Greenhouse area in m^2 and d=internal diameter of the Floating Solar Chimney in m) to be installed in a place of annual horizontal solar irradiation Wy in KWh/m^2 the diagram showing the relation between the annual efficiency of the SAEP and its FSC height H can be calculated.

The following figure (5) shows the annual efficiency as a function of FSC's height for a SAEP of Ac=10^6m^2, d=40m and Wy=1700 KWh/m^2 (Cyprus, South Spain).

The calculated efficiency curve is practically independent of the annual horizontal solar irradiation W_y. However it depends on the FSC internal diameter d. The reason is that a smaller diameter will increase the warm air speed at the top exit of the FSC and consequently will increase the kinetic power losses and decrease the average annual efficiency. If we vary the solar collector diameter of the SAEP its FSC internal diameter should vary proportionally in order to keep almost constant the air speed at the top exit of the FSC and consequently the annual efficiency of the SAEP.

Hence we should notice that in order to receive the efficiency diagram as shown in the following figure (5) figure the kinetic and friction losses of the Floating Solar Chimney should be approximately 10% of the total chimney power. This can be achieved if the internal diameter of the FSC is appropriate in order to keep the average air speed in the range of 7÷8 m/sec, and the FSC internal surface has a low friction loss coefficient.

The following figure (6) shows the variation of the annual efficiency of a SAEP of a FSC 500m high, installed in a place of annual horizontal solar irradiation 1700KWh/m^2 as function of the internal diameter of its FSC.

The annual electricity generated by the SAEP, E_y can be calculated as a product of the annual efficiency and the arriving horizontal solar irradiation on its greenhouse surface $A_c \cdot W_y$. Thus taking into consideration that the annual efficiency is proportional to the FSC

Greenhouse 100 Ha, Annual Horizontal Irradiation1700 KWh/sqm

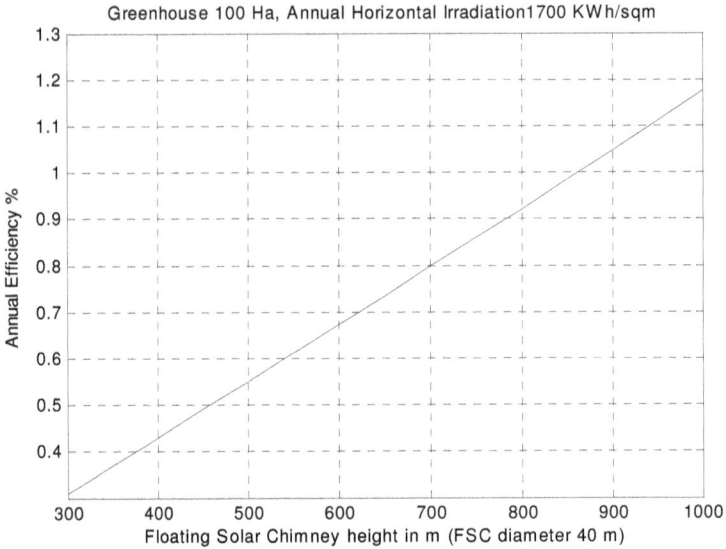

Fig. 5. Annual efficiency of a SAEP as function of its FSC height

height H, the annual generated electricity by the SAEPs is also proportional to the Floating Solar Chimney height H, is as follows:

$$E_y = c \cdot H \cdot A_c \cdot W_y \qquad (4)$$

The constant c is mainly depending on the FSC's internal diameter d.

SAEP of a solar collector 100Ha,in a place with Wy=1700KWh/sqm

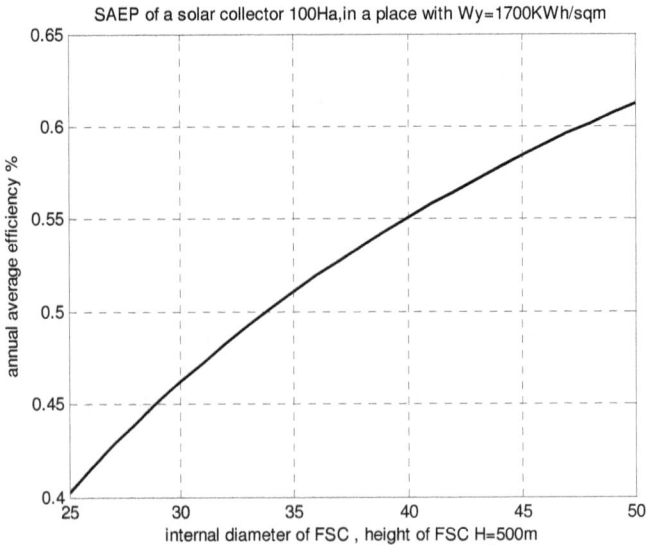

Fig. 6. variation of the annual efficiency of a SAEP with internal FSC diameter

4. Theoretical analysis of the Floating Solar Chimney technology

4.1 Annual average efficiency of SAEPs

The ground thermal storage effect and the daily electricity generation profile, have been studied by several authors (Bernades et.al 2003, Pretorius & Kroger 2006, Pretorious 2007).

The author has used an equivalent approach on the daily power profile study of the floating solar chimney SAEPs using the thermodynamic model see (Backstrom & Gannon 2000) and Fourier series analysis on the time varying temperatures and varying solar irradiance during the 24 hours daily cycle.

Following the code of the author analysis an evaluation of the sensitivity of the various parameters has been made leading to useful results for the initial engineering dimensioning and design of the SAEPs.

The important results of these studies are that the solar chimney power plant annual power production can be increased by using a second glazing below the outer glazing and its output power production can be affected by the ground roughness and ground solar irradiation absorption coefficients.

The thermodynamic cycle analysis proposed in ref. (Gannon Backstrom 2000) is an excellent way of engineering analysis and thermodynamic presentation of the solar chimney power plant operation.

The thermodynamic cycle of the solar chimney operation power plant using the same symbols of the study of ref (Backstrom & Gannon 2000) is shown in the following figure.

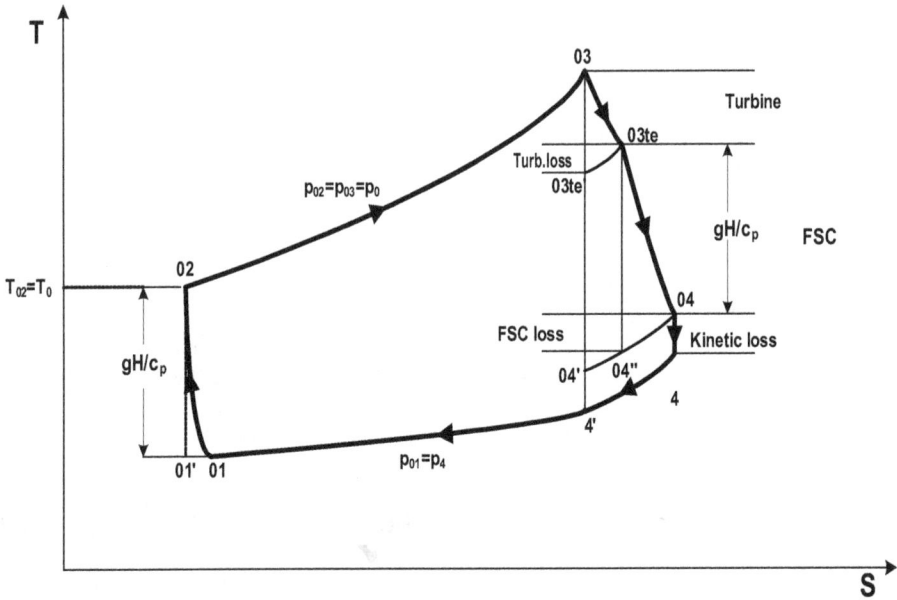

Fig. 7. The thermodynamic diagram of the SAEP

Temperatures, stagnation temperatures (marked with $_0$) and equivalent isentropic temperatures (marked with $'$) are shown in the indicative diagram on the previous figure. The main thermodynamic cycle temperatures are defined in the following table:

T_{01}	Isendropic temperature of ambient air in height H (exit of solar chimney)
T_{02}	Ambient temperature in the ground around the solar collector
T_{03}	Inlet temperature in the air turbines
T_{03te}	Exit air temperature from the turbo generators
T_{04}	Stagnation temperature at the top of the solar collector
T_4	Exit temperature of the air mixed with the ambient air at the top of the exit layers

Table 1.Tthermodynamic cycle temperatures

The process $\{T_{02}$ to $T_{o3}\}$ is assumed as approximately isobaric. This assumption is very reasonable taking into consideration that the heat and expansion of moving air is taking place inside the solar collector.

The processes $\{T_4$ to $T_{01}\}$, $\{T_{03te}$ to $T'_{03te}\}$ and $\{T_{04}$ to $T'_{04}\}$ are definitely isobaric by nature.

By the analysis on the relations between the temperatures the following relationships can be derived:

$$
\begin{cases}
T_{04}=T_4+\dfrac{\alpha \cdot \upsilon_{ex}^2}{2 \cdot c_p}=T_4+C_2 \cdot T_4^2 , T_{03te}=T_{04}+\dfrac{g \cdot H}{c_p} \\[2mm]
T'_{03te}=T_{03}-\dfrac{T_{03}-T_{03te}}{n_T} , T''_{04}=T_{04}-k \cdot \dfrac{\alpha \cdot \upsilon_{ex}^2}{2 \cdot c_p} \\[2mm]
T'_{04}=\dfrac{T'_4 \cdot T_{04}}{T_4} \text{ and } T''_{04}=T_{03} \cdot \dfrac{T'_{04}}{T'_{03te}}
\end{cases}
\tag{5}
$$

Whereby the parameters participating in the relations are defined as follows:

H =solar chimney height

d=internal solar chimney diameter

$A_{ch}=\pi \cdot d^2/4$, is the solar chimney internal cut area

ṁ =moving mass flow

α=kinetic energy correction coefficient, of a usual value of 1.058 calculated in (White 1999).

k=friction loss coefficient inside the solar chimney

k= k_{in}+ 4·C_d ·H/d where, for the operation range of Reynolds numbers inside the solar chimney, the drag friction factor C_d is approximately equal to 0.003, see (White 1999) and for no available data k_{in}it is estimated to 0.15.

η_T= turbo generators overall efficiency, if not available data estimated to 0.8.

T_0= ambient air temperature

T_{02}= T_0+0.5

p_0= ambient atmospheric pressure on ground level at the place of installation of the SAEP, if not available data it is assumed as equal to 101300 Pa.

p_4= ambient atmospheric pressure on top exit at height H, estimated by the formula:

$$
p_4=p_0 \cdot (1-\frac{g \cdot H}{c_p \cdot T_0})^{3.5}
\tag{6}
$$

g= gravity constant 9.81

c_p= specific heat of air approximately equal to 1005
R= air constant approximately equal to 287

v_{ex}= average air speed at the top exit of the solar chimney $\quad v_{ex}=\dfrac{R \cdot T_4 \cdot \dot{m}}{P_4 \cdot A_{ch}}$

and: $\quad T_4'=T_{03} \cdot \dfrac{T_0 \text{-} C_1}{T_0}, C_1=\dfrac{g \cdot H}{c_p}, C_2=\dfrac{a}{2 \cdot c_p} \cdot \left(\dfrac{R \cdot \dot{m}}{A_{ch} \cdot p_4}\right)^2, C_3=T_{03} \cdot (\eta_T \text{-}1)+\dfrac{g \cdot H}{c_p}$

The system of the previous equations can been simplified (see Papageorgiou, 2004), leading to a forth order polynomial equation for T_4 given by:

$$w_1 \cdot T_4^4 + w_2 \cdot T_4^3 + w_3 \cdot T_4^2 + w_4 \cdot T_4 + w_5 = 0 \qquad (7)$$

Where the coefficients w_1, w_2, w_3, w_4 and w_5 are given by the relations:

$$w_1 = C_2^2 \cdot (1\text{-}k), w_2 = C_2 \cdot (2\text{-}k\text{-}\eta_T \cdot C_2 \cdot T_4'), w_3 = (1\text{-}k) \cdot C_2 \cdot C_3 + 1\text{-}2 \cdot \eta_T \cdot C_2 \cdot T_4'$$
$$w_4 = C_3 \text{-} \eta_T \cdot T_4' \cdot (1\text{-}C_1 \cdot C_2), w_5 = \text{-}\eta_T \cdot T_4' \cdot C_1$$

The proper root of the previous polynomial equation is the temperature T_4.
It is easy using the previous relations to calculate T_{03te} by the formula:

$$T_{03te} = T_4 + C_2 \cdot T_4^2 + \frac{g \cdot H}{c_p} \qquad (8)$$

Thus the overall electrical power of the generators is given by the relation:

$$P = \dot{m} \cdot c_p \cdot (T_{03} - T_{03te}) = \dot{m} \cdot c_p \cdot (T_{03} \text{-} T_4 \text{-} C_2 \cdot T_4^2 - \frac{g \cdot H}{c_p}) \qquad (9)$$

As a final result we can say that the air mass flow $\dot{m}$ and the exit temperature T_{03} of the moving air mass through solar collector can define, through the previous analytical procedure, based on the thermodynamic cycle analysis, the electrical power output P of the SAEP.

The proposed thermodynamic analysis, though it looks more complicated than the analysis based on the buoyancy of warm air inside the chimney and the relevant pressure drop to the air turbine used by Bernades M.A. dos S., Vob A., Weinrebe G. and Pretorius J.P., Kroger D.G., it is an equivalent thermodynamic analysis that takes into consideration all necessary and non negligible effects and parameters of the process in the SAEP.

An approximate procedure for T_{03} calculation is given by Shlaigh in his relative book.

The approximate average equation relating the average exit solar collector air temperature T_{03} to its input air temperature T_{02} near the point of optimal operation of the SAEP can be written as follows:

$$ta \cdot G_{av} \cdot A_c = \dot{m} \ C_p \cdot (T_{03} - T_{02}) + \beta \cdot A_c \cdot (T_{03} \text{-} T_{02}) \qquad (10)$$

where:

- β is the approximate thermal power losses coefficient of the Solar Collector (to the ambient and ground) per m2 of its surface area and °C of the temperature difference $(T_{03} \text{-} T_{02})$. An average value of β for double glazing solar collectors is ~3.8÷4 W/m2/°C.

- Gav is the annual average horizontal irradiance on the surface of the solar collector.
- The annual average solar horizontal irradiance G_{av} is given by the formula: $W_y/8760$hours, where W_y is the annual horizontal irradiation of the place of installation of the SAEPP, (in KWh/m²)
- ta is the average value of the product: {roof transmission coefficient for solar irradiation X soil absorption coefficient for solar irradiation}.An average value of the coefficient ta for a double glazing roof is ~ 0.70 .
- and Ac is the Solar Collector's surface area.

Using in the equation an approximation for the function T_{03} ($\dot{m}$), it gives as:

$$T_{03} (\dot{m})= [\text{ta} \cdot G / (\beta + \dot{m} \cdot C_p/A_c)] - T_{02} \tag{11}$$

Where T_{02} is, approximately, equal to the ambient temperature (T_0 in ⁰K), plus 0.5 degrees of Celsius. The increase is due mainly to ground thermal storage around the Solar Collector. The inlet ambient air temperature as passing above it is increasing entering to the solar collector.

The proper value of β, giving the average solar collector thermal losses, has been calculated by the heat transfer analysis of the solar collector. An introduction on this analysis is given on the next paragraph. The heat transfer analysis uses time Fourier series in order to take into account the ground thermal storage phenomena during a daily cycle of operation.

The instantaneous efficiency of the SAEP is given by the formula:

$$\eta = P / (A_C \cdot G) \tag{12}$$

where $A_C \cdot G$ is the solar irradiation power arriving on the horizontal solar collector surface area A_c and P is the maximum generated electric power. This efficiency is for a given value of horizontal solar irradiance G. However we can prove that for an almost constant mass flow near the point of maximum power output, the maximum electric power P and the horizontal irradiance G are almost proportional, thus the previous formula is giving also the annual efficiency of the SAEP defined as the annual generated electricity in KWh divided by the annual horizontal irradiation arriving on top of the roof of the greenhouse of the SAEP i.e

$$\eta = P_{av} / (A_C \cdot G_{av}) = E_y (KWh) / W_Y \tag{13}$$

As an example let us consider that a SAEP has the following dimensions and constants: A_c =10⁶m² (DD=1000m), H = 800 m, d=40 m, k = 0.49, α = 1.1058, η_T = 0.8, the average ambient temperature is T_0 = 296.2 ⁰K and the ambient pressure is P_0 = 101300 Pa. Let us assume that the horizontal solar irradiance G is varying between 100 W/m² to 500 W/m² ($G_{av} \approx$ 240W/m²). In following figure the effect of the G on the power output as function of mass flow of this SAEP is shown.

If the maximum (daily average during summer operation) G_{av} is 500 W/m² the maximum power output of this SAEP, achieved for $\dot{m}_M$ = ~10000 Kg/sec is 5 MW. Thus its efficiency is approximately 1%. Let us assume that the rated power output P_R of a SAEP is the maximum power output for the maximum average solar irradiance. As we can observe on the above figure, the maximum power output point of operation ($\dot{m}_M$) is approximately the same for any horizontal solar irradiance G.

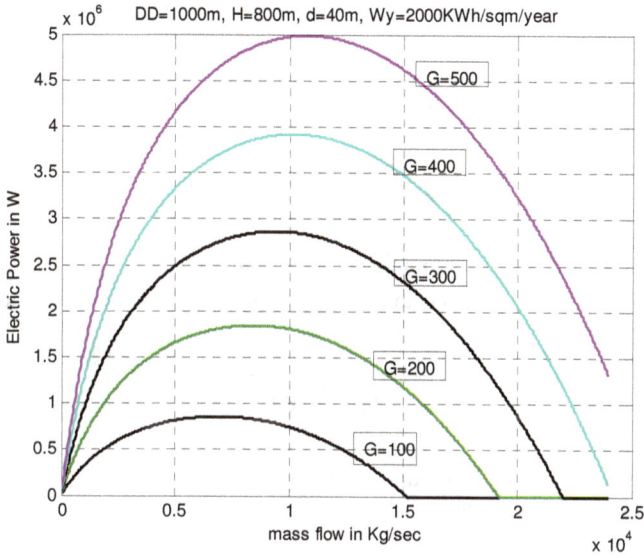

Fig. 9. Electrical Power as function of mass flow for various values of G

Thus if we can control the operation of the SAEP to operate with the proper constant mass flow, close to $\dot{m}_M$, we should achieve almost the possible maximum electric power output by the SAEP for any horizontal solar irradiance. This is referred to as an optimal operation of the SAEP.

As we see later this can be achieved by using induction generators and gear boxes of proper transmission rate.

As a rule of thumb we can state that $\dot{m}_M$ for optimal operation of the SAEP can be calculated approximately by the formula $\dot{m}_M = \rho \cdot \upsilon \cdot (\pi \cdot d^2 / 4)$, where air speed is υ it is estimated to 7-8 m/sec, the air density is given by $\rho = p_0 / (287 \cdot 307.15)$ and d is the internal solar chimney diameter.

A more accurate calculation can be done if we work out on the mass flow for maximum electric power output per annual average horizontal solar irradiance $G_{av,annual} = Wy / 8760$. This can be done using the thermodynamic cycle analysis for variable mass flow $\dot{m}$ and G_{av}. The calculated efficiency for the annual average horizontal solar irradiance $G_{av} = 2100000 / 8760 \approx 240 W / m^2$, of the previously defined SAEP, is 0.94 % (i.e. 6% lower than the calculated efficiency of 1% for the maximum summer average horizontal solar irradiance of $500 W / m^2$).

4.2 Maximum exit warm air speed without air turbines

Using the thermodynamic cycle diagram, the maximum top exit warm air speed of the solar collector plus the FSC alone (i.e. without the air turbines) can be calculated.

In the previous set of equations we should assume that $n_T = 0$. Thus:

$T_{03}=T_{03te}$ and $T_{04}'=T_{04}''$. If we consider that the kinetic losses are approximately equal to $T_{04}'-T_4' \approx \dfrac{a \cdot \upsilon^2}{2 \cdot c_p}$, the friction losses are equal to $T_{04}''-T_{04}' = k \cdot \dfrac{a \cdot \upsilon^2}{2 \cdot c_p}$ and taking into consideration that the equations $T_4' = T_{03} \cdot \dfrac{T_0\text{-}C_1}{T_0}$, $C_1 = \dfrac{g \cdot H}{c_p}$ the following relation is derived:

$$2 \cdot g \cdot H \cdot \frac{\Delta T}{T_0} = (k+1) \cdot a \cdot \upsilon^2 \tag{14}$$

Where $\Delta T = T_{03} - T_0$ (we can approximately consider that $T_0 \approx T_{02}$).

Thus the maximum exit top air speed in a free passage solar chimney (without air turbines) is given by the formula:

$$\upsilon = \sqrt{2 \cdot g \cdot H \cdot \frac{\Delta T}{T_0} / [(k+1) \cdot a]} . \tag{15}$$

For example the exit top speed of the up-drafting air inside the FSC of H=800m height, with ordinary values for coefficients a=1.1058 and k=0.49 and ambient air temperature T_0=296.2 ^{0}K (23 ^{0}C) as function of ΔT is given in the next figure:

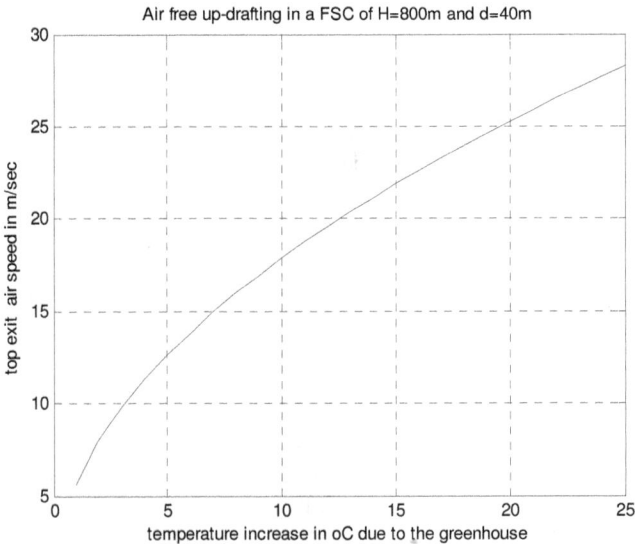

Fig. 10. Free air speed as a function of temperature increase

The temperature increase ΔT as a function of the greenhouse surface area A_c is given by the approximate formula $\Delta T \approx$ ta G / ($\beta + \dot{m}$ c_p/A_c) where ta$\approx$0.7, $\beta \approx$4, c_p=1005, and $\dot{m}_M = \rho \cdot \upsilon \cdot (\pi \cdot d^2/4)$ where $\rho \approx$1.17Kg/m^3, and d=40m. Thus The approximate double glazing solar collector area, generating the free up-drafting air speed υ can be defined by ΔT, υ and G by the equation $A_c \approx \dot{m}$ $c_p/[($ ta G $)/\Delta T - \beta]$.

The approximate solar collector area A_c as a function of the temperature increase ΔT for various values of equivalent horizontal solar irradiance G=250,300,350,400 and 450 W/m², is shown in the following figure.

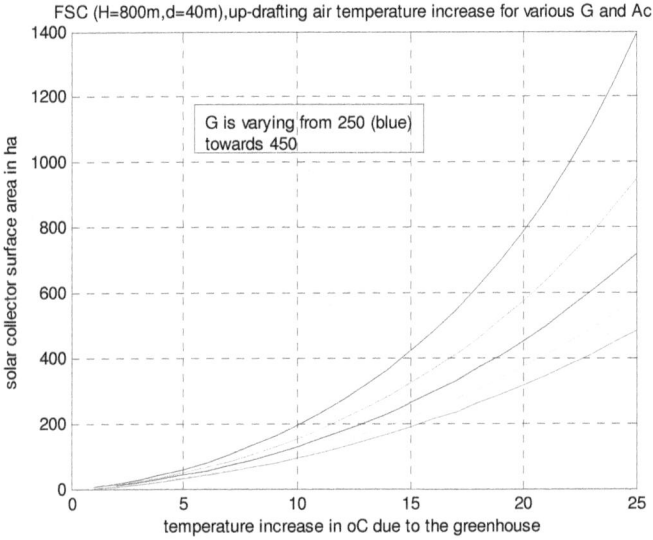

Fig. 11. The solar collector area as a function of its generating temperature increase

Example: for a solar collector of surface area A_c=400Ha (i.e.400000m²), with a diameter $D_c \approx 715m$, for an equivalent horizontal solar irradiation G of 250W/m², the created temperature difference ΔT is ~14.5⁰C and the free up-drafting air speed υ inside the FSC of H=800m height and d=40m internal diameter will be ~21m/sec, while for G=450W/m², ΔT is ~22.5⁰C and υ is ~27m/sec.

For one dimensional analysis $a \approx 1$ and if the friction losses are negligible, i.e. $k \approx 0$, we have:

$$\upsilon \approx \sqrt{2 \cdot g \cdot H \cdot \frac{\Delta T}{T_0}} \qquad (17)$$

Therefore free up-drafting warm air top speed formula, in an adiabatic and free friction FSC, due to its buoyancy, is similar to free falling water speed due to gravity given by:

$$\upsilon_{water} \approx \sqrt{2 \cdot g \cdot H}$$

4.3 The thermal heat transfer model of the SAEP

In order to use the previous thermodynamic cycle analysis of the SAEP we should calculate the warm air temperature T_{03} at the entrance of the air turbine or at the exit of the solar collector. The calculation of this average temperature can be done by using the previously proposed approximate analysis. However the temperature T_{03} is varying during the 24 hours daily cycle.

In order for the daily variation to be calculated and consequently the electric power daily variation using the previously proposed thermodynamic cycle analysis, we should make a heat transfer model and use it for the calculation of the exit temperature as function mainly of daily horizontal irradiance profile and ambient temperature daily profile.

The SAEP heat transfer model with a circular collector is shown in the indicative diagram of the previous figure.

The circular solar collector of this SAEP is divided into a series of M circular sectors of equal width Δr as shown in the next figure.

In this figure the cut of a circular sector of the solar collector of the SAEP is shown with the heat transfer coefficients of the process (radiation and convection) and the temperatures of ground (T_s), moving air (T), inner curtain (T_c), outer glazing (T_w), ambient air (T_0) and sky (T_{sk}). The ground absorbs a part of the transmitted irradiation power due to the horizontal solar irradiance G (ta G).

The wind is moving with a speed υ_w and on the ground it is a thin sheet of water inside a dark plastic film. The ground is characterized by its density ρ_{gr}, its specific heat capacity c_{gr} and its thermal conductivity k_{gr}.

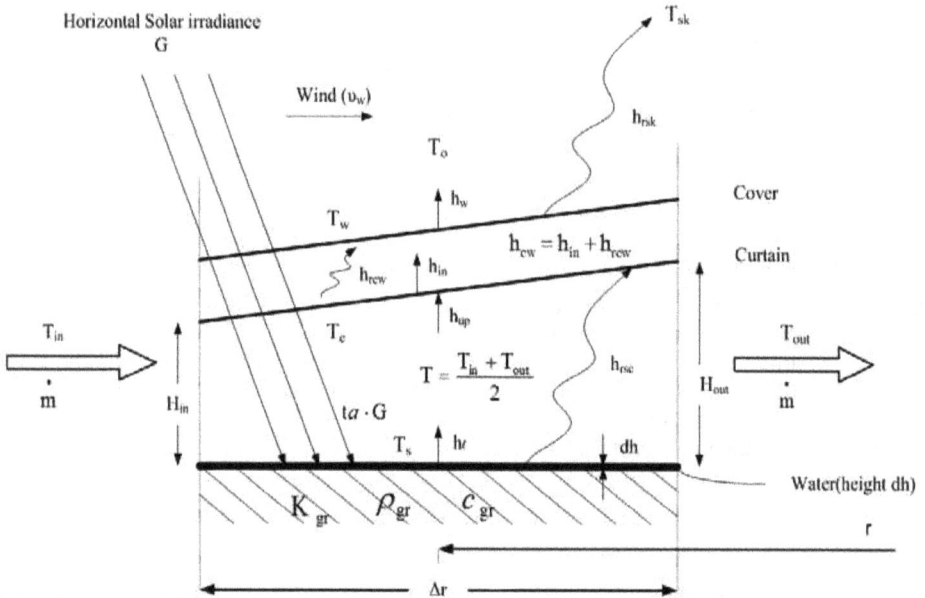

Fig. 12. The cut of a circular sector of a double glazing circular solar collector

The mth circular sector (m=1up to M) will have a width $\Delta r =(D_c-D_{in})/M$, an average radius $r_m= Dc/2-\Delta r \cdot(m1-1/2)$ and an average height $H_m=(H_{in,m}+H_{ex,m})/2$.

For a linear variation of the roof height $H_m= Hin+(Hout-Hin)\cdot (m-1/2)/M$, where D_c=solar collector diameter and D_{in}=Final internal diameter of the solar collector.

These consecutive circular sectors, for the moving air stream of mass flow $\dot{m}$, are special tubes of nearly parallel flat surfaces and therefore they have equivalent average diameters $d_{e,m}=2*H_m$.

As the ambient air moves towards the entrance of the first circular sector it is assumed that its temperature T_0 increases to T_0+dT due to the ground heat transfer convection to inlet air, around the solar collector. As an approximation dT is estimated to 0.5 °K.

The exit temperature of the first sector is the inlet temperature for the second etc. and finally the exit temperature of the final M^{th} sector is the T_{03}, i.e. the inlet stagnation temperature to the air turbines.

The solar chimney heat transfer analysis during a daily 24 hours cycle, is too complicated to be presented analytically in this text however we can use the results of this analysis in order to have a clear picture of the operational characteristics of the SAEPs. Using the code of the heat transfer analysis for moving mass flow $\dot{m}_M$, the daily variation of the exit temperature T_{03} can be calculated. Using these calculated daily values of the T_{03} and by the thermodynamic cycle analysis for the optimal mass flow $\dot{m}_M$ the daily power profile of the electricity generation can be calculated.

With this procedure the 24 hour electricity generation power profile of a SAEP with a solar collector of surface area $A_c=10^6 m^2$ and a FSC of H=800m height and d=40m internal diameter for an average day of the year has been calculated. The SAEP is installed in a place with annual horizontal solar irradiation W_y=1700 KWh/m².

In the following figure three electric power profiles are shown with or without artificial thermal storage.

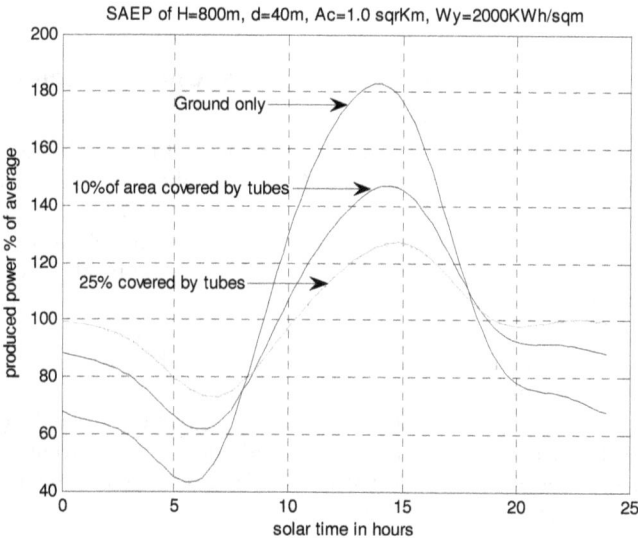

Fig. 13. The average daily SAEP's electricity generating profiles

The relatively smooth profile shows the electric power generation when only the ground acts as a thermal storage means. While the smoother profiles are achieved when the greenhouse is partly covered (~10% or ~25% of its area) by plastic black tubes of 35cm of diameter filled with water, i.e. there is also additional thermal storage of an equivalent water sheet of $35 \cdot \pi/4 = 27.5$ cm on a small part of the solar collector.

The daily profiles show that the SAEP operates 24hours/day, due to the greenhouse ground (and artificial) thermal storage. That is a considerable benefit of the FSC technology compared to the rest solar technologies and the wind technology which if they are not equipped with energy mass storage systems they can not operate continuously.

As shown in the produced curves on the previous figure, with a limited (~10%) of the greenhouse ground covered by plastic tubes (35 cm) filled with water, the maximum daily power is approximately 140% of its daily average, or the daily average is 70 % of its maximum power.

Taking into consideration the seasonal power alteration and assuming that the average annual daily irradiation at a typical place is approximately 70% of the average summer daily irradiation, the annual average power can be estimated as a percentage of the maximum power production (at noon of summertime) as the product of 0.77·0.70=0.49.

The maximum power is equal to the rating of the power units of the SAEP (Air turbine, electric generator, electric transformer etc.), while the average power multiplied by 8760 hours of the year defines the annual electricity generation. Therefore the capacity factor of a SAEP equipped with a moderate artificial thermal storage can be as high as ~49%.

Without any artificial thermal storage the average daily power is approximately 0.55 of its maximum thus the capacity factor is ~37% (0.55·0.70≈0.385).

This means that in order to find the annual energy production by the SAEP we should multiply its rating power by ~3250÷4300 hours. However we should take into consideration that the SAEPs are operating continuously (24x365) following a daily and seasonal varying profile.

5. The major parts and engines of Floating Solar Chimney technology

5.1 The solar collector (Greenhouse)

The solar collector can be an ordinary circular greenhouse with a double glazing transparent roof supported a few meters above the ground. The periphery of the circular greenhouse should be open to the ambient air. The outer height of the greenhouse should be at least 2 meters tall in order to permit the entrance of maintenance personnel inside the greenhouse. The height of the solar collector should be increased as we approach its centre where the FSC is placed. As a general rule the height of the transparent roof should be inversely proportional to the local diameter of the circular solar collector in order to keep relatively constant the moving air speed. The circular greenhouse periphery open surface can be equal or bigger than the FSC cut area.

Another proposal with a simpler structure and shape the greenhouse can be of a rectangular shape of side DD. The transparent roof could be made of four equal triangular transparent roofs, elevating from their open sides towards the centre of the rectangle, where the FSC is placed. Thus the greenhouse forms a rectangular pyramid.

The previous analysis is approximately correct and can be figured out by using an equivalent circular greenhouse external diameter $D_c \approx DD \cdot \sqrt{4/\pi}$.

The local height of each inclined triangular roof is almost inversely proportional to the local side of the triangle in order to secure constant air speed.

Both solar collector structures are typical copies of ordinary agriculture greenhouses although they are used mainly for warming the moving stream of air from their periphery towards the centre where the FSC of the SAEP is standing. Such greenhouses are appropriate for FSC technology application combined with special agriculture inside them.

In desert application of the FSC technology the solar collectors are used exclusively for air warming. Also in desert or semi desert areas the dust on top of the transparent roofs of the conventional greenhouses could be a major problem. The dust can deteriorate the transparency of the upper glazing and furthermore can add unpredictable weight burden on the roof structure. The cleaning of the roof with water or air is a difficult task that can eliminate the desert potential of the FSC technology.

Furthermore in desert or semi-desert areas the construction cost of the conventional solar collector (a conventional greenhouse) could be unpredictably expensive due to the unfavourable working conditions on desert sites.

For all above reasons another patented design of the solar collectors has been proposed by the author.The proposed modular solar collector, as has been named by the author, will be evident by its description that it is a low cost alternative solar collector of the circular or rectangular conventional greenhouse which can minimize the works of its construction and maintenance cost on site.

We can also use and follow the ground elevation on site, and put the FSC on the upper part of the land-field therefore the works on site for initial land preparation will be minimized.

The greenhouse will be constructed as a set of parallel reverse-V transparent tunnels made of glass panels as shown in the next figure (14). The maximum height of the air tunnel should be at least 190cm in order to facilitate the necessary works inside the tunnel, as it is for example the hanging of the inner crystal clear curtains.

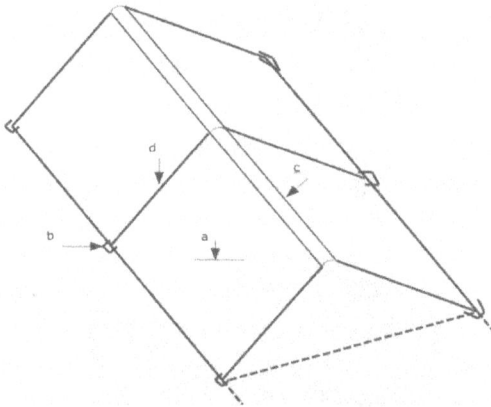

Fig. 14. A part of the triangular tunnel of two panels (a)glass panel, (b)ground support, (c)glass panel connector (d)glass plastic separator

An indicative figure of a greenhouse made of ten air tunnels is shown in next figure. Among the parallel air tunnels it is advisable that room should be made for a corridor of 30-40cm of width for maintenance purposes.

By above description it is evident that the modular solar collector is a low cost alternative of a conventional circular greenhouse for the FSC technology in desert or semi-desert areas that minimize the works on site and lower the construction costs of the solar collector and its SAEP. Furthermore the dust problem is not in existence because the dust slips down on the inclined triangular glass panels.

The average annual efficiency of the modular solar collector made by a series of triangular warming air tunnels with double glazing transparent roofs is estimated to be even higher than 50%. Thus its annual efficiency will follow the usual diagram of efficiency (or it will be even higher).

The total cut area of all the triangular air tunnels should be approximately equal to the cut area of the FSC for constant air speed. The central air collecting corridor cut should also follow the constant air speed rule for optimum operation and minimum construction cost.

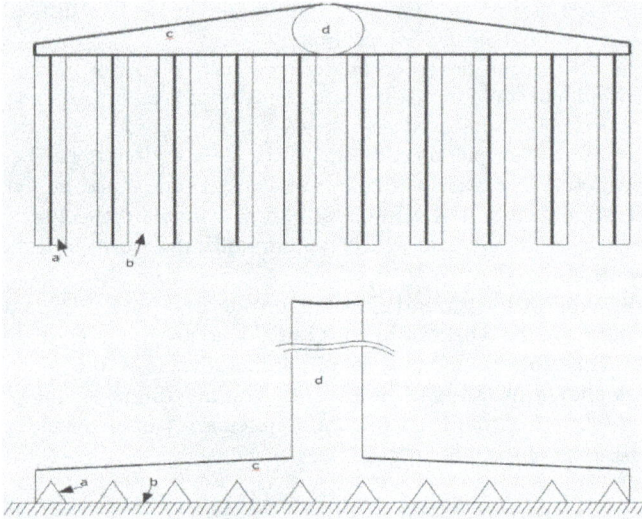

Fig. 15. Modular solar collector with ten air tunnels (a)Triangular tunnel, (b)Maintenance corridor (c)Central air collecting tube, (d)FSC

5.2 The Floating Solar Chimney (FSC)
A small part of a typical version of the FSC on its seat is taking place in the figure(16) below.

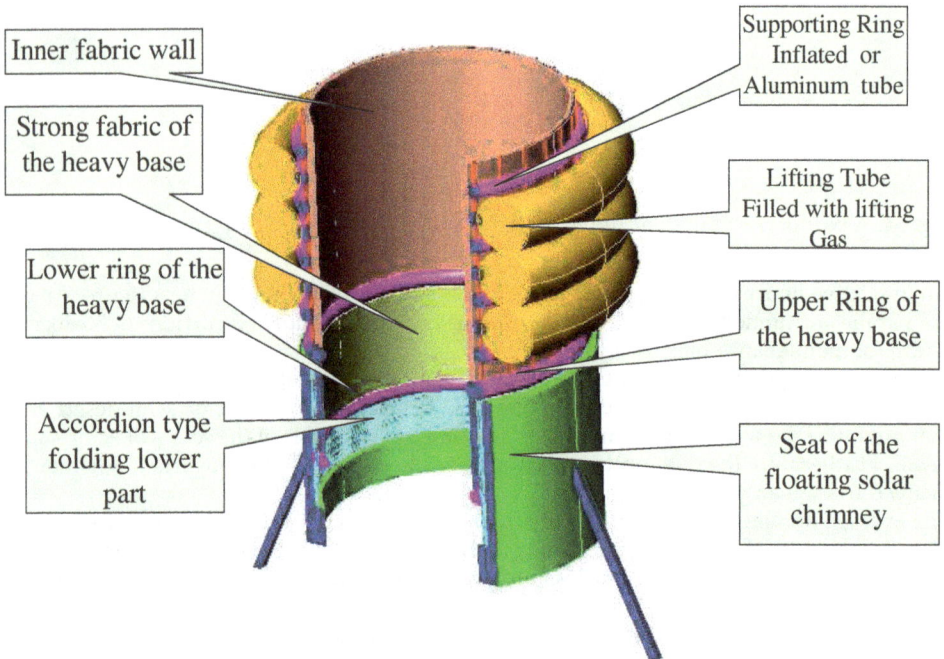

Inner fabric wall

Strong fabric of the heavy base

Lower ring of the heavy base

Accordion type folding lower part

Supporting Ring Inflated or Aluminum tube

Lifting Tube Filled with lifting Gas

Upper Ring of the heavy base

Seat of the floating solar chimney

Fig. 16. A small part of a typical version of the FSC on its seat

The over-pressed air tubes of the fabric structure retain its cylindrical shape. While the lifting tubes (usually filled with NH_3) supply the structure with buoyancy in order to take its upright position without external winds. Both tubes can be placed outside the fabric wall as they are shown in the figure or inside the fabric wall. When the tubes are inside the fabric core they are protected by the UV radiation and the structure has a more compact form for the encountering of the external winds unpredictable behavior. But inside the warm air friction losses are increased and in order to have the same internal diameter the external diameter of the fabric core should be greater. In the first demonstration project both shapes could be tested in order that the best option is chosen.

Therefore the FSCs of the SAEPs are free standing fabric structures and due to their inclining ability they can encounter the external winds. See the next indicative figure (17) describing its tilting operation under external winds.

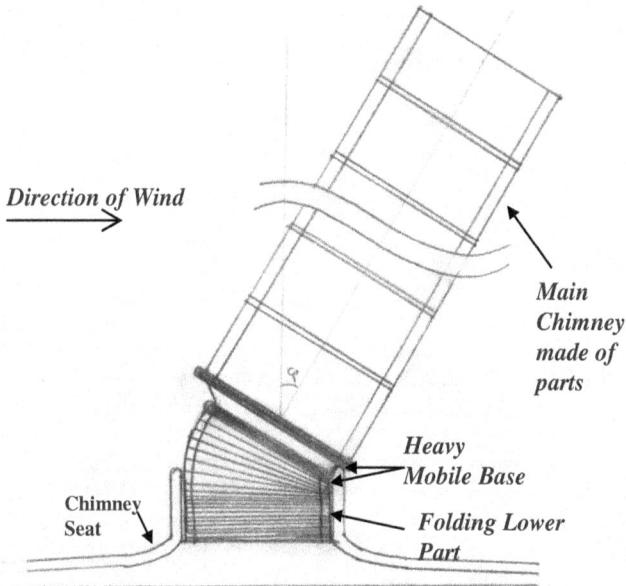

Fig. 17. Tilting operation of the FSC under external winds

However in areas with annual average strong winds the operating heights of the inclining fabric structures are decreasing. The following figure (18) presents the operating height loss of the FSCs as function of the average annual wind speed, for Weibull average constant $k \approx 2.0$. The net buoyancy of the FSC is such that will decline 60^0 degrees when a wind speed of 10 m/sec appears.

For example using the diagram in figure (18), for an average wind speed of 3 m/sec and a net lift force assuring a 50% bending for a wind speed of 10 m/sec, the average operating height decrease is only 3.7%.

As a result we can state that the best places for FSC technology application are the places of high average horizontal solar irradiation, low average winds and limited strong winds. The mid-latitude desert and semi-desert areas, that exist in all continents, combine all these properties and are excellent places for large scale FSC technology application.

weibull constant k=2; decline 50 % for v=10 m/sec

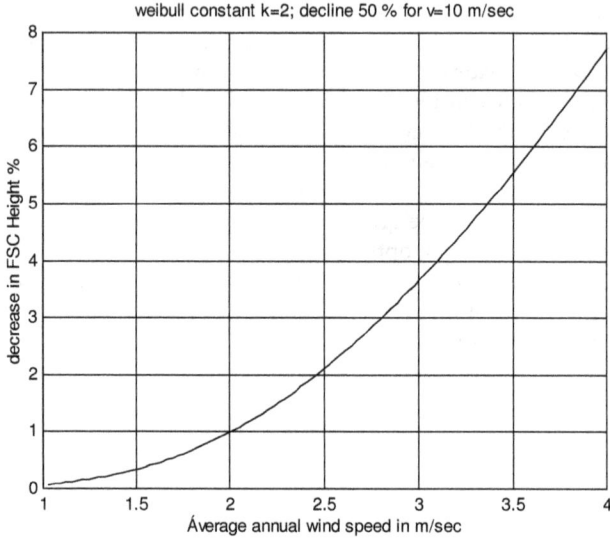

Fig. 18. FSC's operating height average decrease under external winds.

5.3 The air turbines

The air turbines of the SAEPs are either of horizontal axis placed in a circular pattern around their FSCs or with normal axis placed inside the FSCs (near the bottom). The later case with only one air turbine is most appropriate for the FSC technology, while the former is more advisable for concrete solar chimney technology applications.

The air turbines of the solar chimney technology are caged (or ducted) air turbines. These air turbines are not similar to wind turbines that transform the air kinetic energy to rotational energy, therefore their rotational power output depends on the wind speed or the air mass flow. The caged air turbines transform the dynamic energy of the warm air, due to their buoyancy, to rotational. Therefore their rotational power output does not depend on the mass flow only but on the product of the mass flow and the pressure drop on the air turbine. Therefore the warm air mass flow, as we have noticed already, is possible to remain approximately constant during the daily operation (in order that an optimal operation is achieved) while its rotational power and its relative electric power output vary during the daily cycle. The varying quantity is the pressure drop of the air turbine. This pressure drop depends on the warm air temperature i.e. the warm air proportional buoyancy and the FSC height.

The air turbines are classified according to the relation between their mass flows and their pressure drops. The wind turbines are class A turbines (large mass flow small pressure drop). The useful classes for solar chimney application are the class B and C. The class B are the caged air turbines with lower pressure drop and relatively higher mass flow and made without inlet guiding vanes, while the class C air turbines are with higher pressure drops and relatively lower mass flows and should be made of inlet guiding vanes in order that optimal efficiency is achieved.

Considering that the floating or concrete solar chimney SAEPs can have the same heights (between 500m÷1000m) the defining factor for air turbines with or without inlet guiding vanes is the solar collector diameter.

For the expensive concrete solar chimney the respective solar collectors are made with high diameters in order to minimize the construction cost of their SAEPs. While the low cost floating solar chimneys can be designed with smaller solar collectors for minimal cost and optimal operation.

The diameters of the solar collectors are proportional to the increase of the warm air temperatures $\Delta T = T_{03} - T_0$, thus proportional also to the buoyancies and to the pressure drops on the air turbines.

Therefore the Floating Solar Chimney SAEPs can be designed with air turbines of class B (i.e. without inlet guiding vanes). These caged air turbines are lower cost units per generated electricity KWh in comparison with class C air turbines which are appropriate for concrete solar chimney SAEPs.

5.4 The electric generators

There are two types or electric generators which can be used in SAEPs, the synchronous and the induction or asynchronous electric generators.

The synchronous electric generators for FSC technology should have a large number of pole-pairs pp. The frequency of the generated electricity by the multi-pole synchronous electric generator should be equal to the grid frequency f.

The generated electricity frequency of the synchronous generators f_{el} is proportional to its rotational frequency f_g i.e. $f_{el} = pp \cdot f_g$. Thus in case of varying f_g an electronic drive is necessary, for adjusting the generated electric frequency f_{el} to the grid electric frequency f.

A multi-pole (high value of pp) synchronous electric generator combined with an electronic drive can be a reasonable solution in order to avoid the adjusting gear box.

In order to control the set to operate the whole SAEP under optimal conditions we either control its electronic drive unit or its air turbine blade pitch.

The induction generators are of two types. The squirrel cage and the double fed or wound rotor induction generators. The squirrel cage induction generators rotate with frequencies close to their synchronous respective frequencies f/pp defined by the grid frequency and their pole-pairs. For given pole-pairs (for example for four pole caged induction generators pp=2) the induction generator should engage itself to the air turbine through an appropriate gear box that is multiplying its rotational frequency in order that the generator rotational speed matches to the frequency $(f/pp) \cdot (1+s)$, where s is the absolute value of the slip and it is a small quantity in the range of 0.01 for large generators.

The electric power output of the squirrel cage induction generator is approximately proportional to the absolute value of the slip s near their operating point. Thus even high power variations can be absorbed with small rotational frequency variations. Therefore the squirrel cage induction generators engaged to the air turbines with proper gear boxes are supplying the grid always with the proper electric frequency and voltage without any electronic control. The only disadvantage of the squirrel cage induction generators is that they always produce an inductive reactive power. This reactive power should be compensated using a parallel set of capacitors creating a capacitive reactive power.

The wound rotor or doubly fed induction generators are characterized by the fact that their rotors are supplied with a low frequency electric current. With proper control of the voltage and frequency of the rotor supply we can make them operate as zero reactive power units. The electronic system supplying the rotor with low frequency current is a power electronic unit of small power output (~3% of the power output of the generator). However the doubly fed induction generators with these small electronic supplies of their rotors are more expensive than the squirrel cage induction generators with reactive power compensating capacitors.

The SAEPs with normal axis air turbines have enough space underneath the air turbine to accommodate a large diameter multi-pole generator with a large number of pole pairs in order to avoid the rotation frequency adjusting gear box.

I believe that the large scale application of the FSC technology will boost the research and production of large diameter multi-pole squirrel caged or wound rotor induction generators in order to avoid the sensitive and expensive adjusting gear boxes and to lower the cost of large electronic drives of multi-pole synchronous generators.

5.5 The gear boxes

The gear box is a essential device for adjusting the frequency of the rotation of the air turbines f_T to the electric frequency f of the grid through the relation

$f = pp \cdot f_T \cdot rt$. The rt is the rate of transmission of the gear box i.e the generator rotates with frequency $f_g = f_T \cdot rt$.

When conventional electric generators with a few pole pairs (low pp) are used, as electricity generating units, gear boxes with a proper rate of transmission rt are necessary. However if multi-pole electric generators are used with high pole-pair values (pp_h) then the gear boxes can be avoided (if $pp_h = pp \cdot rt$).

The gear boxes are mechanical devices made of gears of various diameters and combinations in order to transform their the mechanical rotation incoming and out-coming characteristics (i.e.the frequency of rotation f_{in}, f_{out} and the torque Tq_{in} and Tq_{out}) by the relations $f_{in}/f_{out} = Tq_{out}/Tq_{in} = rt =$ rate of transmission.

The gears demand a continuous oil supply and have a limited life cycle. Thus the gear boxes being huge and heavy devices of high maintenance and sensitivity, if possible they should not be preferred.

The electric power production by the SAEPs, is calculated as a function of the inlet air speed υ (i.e. the air mass $\dot{m}$) in the air turbines by a relation of the form:

$$P = \dot{m} \cdot c_p \cdot \left(T_{03} - T_{03te}\right) = \dot{m} \cdot c_p \cdot \left(T_{03} - T_4 - C_2 \cdot T_4^2 - \frac{g \cdot H}{c_p}\right) \tag{9}$$

Where T_{O3}, T_{O3te} are functions of mass flow $\dot{m}$ and FSC top exit temperature T_4.

We have shown that T_4 is the (appropriate) root of a fourth order polynomial equation:

$$w_1 \cdot T_4^4 + w_2 \cdot T_4^3 + w_3 \cdot T_4^2 + w_4 \cdot T_4 + w_5 = 0 \tag{7}$$

where w_1, w_2, w_3, w_4 and w_5 are functions of the geometrical, the thermal and ambient parameters of the SAEP, the air turbine efficiency η_T and the equivalent horizontal solar irradiance G.

The mass flow $\dot{m}$ and the warm air speed υ are proportional ($\dot{m} = \rho \cdot A_t \cdot \upsilon$) Thus:

P=Function (υ)

The efficiency of the air turbine is in general a function of the ratio υ / υ_{tip} i.e. $\eta_T(\upsilon$ / $\upsilon_{tip})$ where υ_{tip} is the blades' end rotational speed.

The air turbines of the SAEPs with their geared electric generators are generating electric power following the air turbine characteristics given by the two operating functions P (υ), and η_T (υ / υ_{tip}). Considering that $\upsilon_{tip} = \pi \cdot f_T \cdot d_T$, where f_T is the air turbine frequency of rotation and d_T the turbine diameter.

The electric frequency for the geared electric generators is equal to f_n where: $f_n = f_t \cdot rt \cdot pp$, rt is the gear box transmission ratio and pp the number of their pole pairs. Hence:

$$\upsilon_{tip} = \frac{\pi \cdot d_T \cdot f_n}{rt \cdot pp} \qquad (18)$$

For optimal power production by a SAEP, for an average solar irradiance G, the maximum point of operation of P(υ) should be reached for an air speed υ for which the efficiency η_T (υ / υ_{tip}) is also maximum.

The value of υ_m for maximum electric power can be defined by the SAEP operating function for η_T=constant (usually equal to 0.8) and a given solar irradiance G.

The value of the ratio (υ / υ_{tip})$_m$ for maximum air turbine efficiency can be defined by the turbine efficiency function η_T(υ / υ_{tip}).

Thus the appropriate υ_{tip} is defined by the relation:

$$\upsilon_{tip,m} = \frac{\upsilon_m}{\left(\upsilon \middle/ \upsilon_{tip} \right)_m} \qquad (19)$$

Where the index m means maximum power or efficiency.

Thus for $\upsilon_{tip,m}$ the maximum power production under the given horizontal solar irradiance G is generated. Taking into account that υ_{tip} and f_n are proportional, f_n should vary with the horizontal solar irradiance G.

However as we have stated the mass flow for maximum power output by the SAEP is slightly varying with varying G, thus we can arrange the optimum control of the SAEP for the average value of G.

A good choice for this average G is a value of 5÷10% higher than the annual average $G_{y,av}$, defined by the relation $G_{y,av}=W_y/8760$.

Following the previous procedure for the proposed G, if the air turbine efficiency function η_T(υ / υ_{tip}) is known or can be estimated, the value of $\upsilon_{tip,m}$ can be calculated.

The frequency f of the produced A.C. will follow f_n by the relation f = $(1+s) \cdot f_n$, where s is the absolute value of the operating slip. Taking into consideration that the absolute value of slip s, for large induction generators, is less than 1%, f≈f_n.

Thus the gear box transmission ratio will be defined by the approximate relation:

$$rt \approx \frac{\pi \cdot d_T \cdot f}{\upsilon_{tip,m} \cdot pp} \qquad (20)$$

If the air turbine efficiency function η_T(υ / υ_{tip}) is not known we can assume that for caged air turbines without inlet guiding vanes their maximum efficiency is achieved for $\upsilon_{tip,m}=(6÷8) \cdot \upsilon$.

Thus:

$$rt \approx \frac{\pi \cdot d_T \cdot f}{(6 \cdots 8) \cdot \upsilon_m \cdot pp} \qquad (21)$$

Where: υ_m= the air speed for maximum efficiency of the SAEP (derived by the SAEP basic equation for the chosen value of G), d_T= the caged air turbine diameter (smaller by 10% of the FSC diameter usually), f=the grid frequency (usually 50 sec^{-1}), pp=2 (usually the generators are four pole machines).

6. Dimensioning and construction cost of the Floating Solar Chimney SAEPs

6.1 Initial dimensioning of Floating Solar Chimney SAEPs

The floating solar chimneys are fabric structures free standing due to their lifting balloon tube rings filled with a lighter than air gas. The inexpensive NH_3 is the best choice as lifting gas for the FSCs. As we will see later the FSCs are low cost structures, in comparison with the respective concrete solar chimneys.

The annual electricity generation by the SAEPs (E) is proportional to their FSC's height (H), their solar collector surface area (A_c) and the annual horizontal irradiation at the place of their installation W_y i.e. $E=c\cdot H\cdot A_c\cdot W_y$.

As for the concrete solar chimney SAEPs, due to their concrete solar chimneys high cost, it is obvious that in order to minimize their overall construction cost per produced KWh, it is preferable to use one solar chimney, of height H and internal diameter d, and a large solar collector of surface area A_c.

In case of the floating solar chimney SAEPs, generating the same annual amount of electricity, a farm of N similar SAEPs should be used. Their FSCs will have the same height (H) and their solar collectors a surface area A_c/N. If the internal diameters of these FSCs are $d_{FSC} \approx d/\sqrt{N}$ then both Power Plants they will have the same efficiency and power production. Usually $d_{FSC} > d/\sqrt{N}$ therefore the FSC farm has higher efficiency and generates more electricity than the concrete solar chimney SAEP for the same solar collector area.

We have several benefits by using farms of FSC technology as for example:

- The handling of FSC lighter than air fabric structures is easy if their diameters are smaller. The diameter d_{FSC} should not be less than 1/20 of FSC height H.
- This choice will give us the benefit of using existing equipment (electric generators, gear-boxes, etc.) already developed for the wind industry.
- The smaller surface areas of the solar collectors will decrease the average temperature increase ΔT of the moving air mass, and consequently it is advisable that simpler and lower cost air turbines should be used (class B instead of class C air turbines i.e. caged air turbines without inlet guiding vanes).

The following restrictions are prerequisite for a proper dimensioning of the Floating Solar Chimney SAEPs.

- The FSC height H should be less than 800m.
- Their internal diameter should be less than 40m
- The solar collector active area should be less than 100 Ha (i.e. $10^6 m^2$)

If the solar collectors are equipped with artificial thermal storage the SAEP will have a rating power of $P_r=W_y\cdot\eta\cdot A_c/4300$. For maximum height 800m, and d=40m the SAEP annual efficiency is $\eta\approx1\%$. In desert places W_y can be as high as 2300 KWh/m^2. Thus P_r for the maximum solar collector surface area of $10^6 m^2$ is less than 5MW.Generators and respective gear-boxes up to 5MW are already in use for wind technology. Furthermore if we choose an internal diameter of 40m for the FSC, it can be proven that for rating power less than 5MW, the optimal air turbine should be of class B, i.e. without the inlet guiding vanes. The air turbine will be placed onto the normal axis inside the bottom of the FSC. A useful notice concerning the dimensioning of the SAEPs is that for constant FSC height H, rating power and annual horizontal irradiation the solar collector equivalent diameter D_c and the FSC internal diameter d are nearly proportional.Let us apply the dimensioning rules in the case of desert SAEPs, considering for example that the annual horizontal irradiation is not less than 2100 KWh/m². Let us consider that the FSC height H is varying, while the solar collector area is remaining constant to 1.0Km² and the FSC internal diameter is also constant and equal to 40m. The rating power of the respective SAEPs, with artificial thermal storage, is shown on the following table(2).

Solar collector area in Km²	FSC internal diameter d in m	FSC height H in m	Rating power P_r in MW
1.0	40	180	1.0
1.0	40	360	2.0
1.0	40	540	3.0
1.0	40	720	4.0
1.0	40	800	4.5

Table 2. Dimensions and rating of SAEPs of 1Km² with artificial thermal storage

In the following table (3) initial dimensions of the SAEPs of FSC height 720m installed on the same area for rating power 1MW, 2MW, 3MW and 4 MW are shown.

Solar collector area in Km²	Minimum FSC internal diameter d in m	FSC height H in m	Rating power P_r in MW
0.25	36	720	1.0
0.50	36	720	2.0
0.75	36	720	3.0
1.0	36	720	4.0

Table 3. Dimensions and rating of SAEPs of 720m height with artificial thermal storage

6.2 Estimating the direct construction cost of Floating Solar Chimney SAEPs

The direct construction cost of a Floating Solar Chimney SAEP with given dimensions is the sum of the costs of its three major parts, the solar collector cost (C_{SC}), the FSC cost (C_{FSC}) and the Air turbines gear boxes and generators cost (C_{TG}).The construction cost of the solar collector is proportional to its surface area. A reasonable rough estimate of modular solar collectors including the cost of their collecting corridors is:

$$C_{SC}=6.0 \cdot A_c \text{ in EURO } (A_c \text{ in } m^2) \tag{22}$$

The construction cost of the FSC is the sum of the cost of its fabric lighter than air cylinder, and the cost of the heavy base, the folding accordion and the seat. A reasonable rough estimation of above costs is:

$$C_{FSC}=60 \cdot H \cdot d + 300 \cdot d^2 \text{ in EURO (H, d in m)} \tag{23}$$

The construction cost of the Turbo-Generators is proportional to the rating power P_r of the SAEP a reasonable rough estimation for this cost is:

$$C_{TG}=300 \cdot P_r \text{ in EURO } (P_r \text{ in KW}) \tag{24}$$

The estimating rough figures are reasonable for SAEPs of rating power of $1 \div 5$ MW. Any demonstration SAEP and maybe the first few operating SAEPs possible will give us a construction cost up to ~100% higher than the estimated by the previous rough formulae but gradually the direct construction cost of the SAEPs should have even lower construction costs than estimated by the given rough formulae. In the following tables (4,5) the construction costs of the previously dimensioned SAEPs are given.

Taking into consideration that the rating power multiplied by 4300 hours (for solar collectors reinforced with artificial thermal storage) will give the annual electricity generation, the construction cost per produced KWh/year is also presented in the tables (4,5).

Solar collector area in Km²	FSC internal diameter d in m	FSC height H in m	Rating power P_r in MW	Construction cost in million EURO	Construction cost in EURO per produced KWh/year
1.0	40	180	1.0	7.2	1.54
1.0	40	360	2.0	8.0	0.85
1.0	40	540	3.0	8.7	0.62
1.0	40	720	4.0	9.4	0.50
1.0	40	800	4.5	9.8	0.47

Table 4. Direct construction cost of various SAEPs

Solar collector area in Km²	Minimum FSC internal diameter d in m	FSC height H in m	Rating power P_r in MW	Construction cost in million EURO	Construction cost in EURO per produced KWh/year
0.25	36	720	1.0	2.75	0.64
0.50	36	720	2.0	5.45	0.63
0.75	36	720	3.0	7.35	0.57
1.0	36	720	4.0	9.15	0.53

Table 5. Direct construction cost of various SAEPs

7. Floating Solar Chimney versus concrete chimney SAEPs

The optimum dimensions and power ratings of the concrete solar chimney SAEPs are far higher than the Floating Solar Chimney dimensions and rating. In order for them to be compared we should consider a concrete solar chimney SAEP with given dimensions and construction cost and a Floating Solar Chimney SAEP farm generating annually the same electricity and having the same solar chimney height.

In a paper presented in 2005 (Shlaigh et al., 2005) it was mentioned the estimates on the construction cost of large SAEPs of concrete solar chimneys (Solar Updrafts Towers as they name them). According to these estimates concerning a 30 MW SAEP with a concrete solar chimney of 750 m height and 70 m of internal diameter and a solar collector of 2900m diameter(i.e. 6.6 Km² of surface area) the SAEP will generate 99 million KWh/year and will have a construction cost of 145 million EURO (2005 prices). Prof Jorg Schlaigh in a recent speech was estimating the construction cost of a similar concrete solar chimney SAEP of a solar chimney of 750m height and 3Km diameter to be 250÷300 million EURO (prices 2010).

Let us compare this concrete chimney SAEP with a farm of 9 Floating Solar Chimney SAEPs each one with a solar collector of surface area 740000m² (all of them together will cover approximately the same land area of the concrete solar chimney SAEP of 6.6Km²). Furthermore let as assume that all of them have the same FSC of ~750m height and an internal diameter of ~40m. Let us also assume that the power rating of each FSC SAEP is ~3MW.

Although it is reasonable to assume that with these assumptions both electricity generating power plants will generate the same KWh of electricity per year (~99million KWh/year), the FSC farm could generate30% more electricity. This is the result of having a higher overall solar chimney cut in the farm of nine SAEPs, or equivalently the FSC farm will have an equivalent solar chimney diameter of 120m ($120m = 40m \cdot \sqrt{9(SAEPs)}$). Thus the warm air speed, in the FSCs, is lower than the air speed within the concrete chimney, therefore the kinetic energy losses of the exit air are lower in the FSCs and the efficiency of the FSC farm is higher.

Using the previous construction cost relations the estimated construction cost of each Floating Solar Chimney SAEP of the farm is ~6million EURO (2010 prices). Thus the whole FSC farm will have a construction cost of 54 million EURO.

The final result is that the capital expenditure for the Floating Solar Chimney farm, for similar electricity generation with the concrete solar chimney solar updraft tower, is 3 to 5 times smaller.

8. Direct production cost of electricity KWh of the FSC technology

8.1 Direct production cost analysis

The direct production cost of MWh of any electricity generating power plant is the sum of three costs:

- The capital cost related to the capital expenditure (CapEx) on investment
- The operation and maintenance cost
- The fuel cost
- The CO_2 emission cost

For renewable technology PPs the fuel and the carbon dioxide emission costs are zero.

The base load continuous operating technologies are dominating the electricity generation and their average estimated direct production cost per MWh is, without any carbon emission penalty within the range of 55÷60 EURO (EU area 2009).

The onshore wind turbine farms have succeeded to generate electricity almost with the same cost in average. However it is generating intermittent electricity thus it can enter to the grid up to 45% in power and cover the 15÷20 of the electricity demand.

Let us calculate the direct production cost of the solar chimney technology.

The assumptions we use are the following for FSC and concrete solar chimney SAEPs:

- The life cycle of both SAEPs is high (minimum 40 years)

- The CapEx is a long term loan repaid in 40 equal installments
- The interest rate of above loans is 6% (2009)
- The fabric FSCs should be replaced every 6÷10 years. This cost goes along with the maintenance cost.
- The initial construction period of the concrete chimney SAEPs is 3÷5 years while the period for FSC SAEPs is 1÷2 years. The repayments will start after those periods.
- Thus the annual repayment installment will be equal to 7% for the FSC farm and 7.5% for the concrete solar chimney PP (with the cost of initial grace period to be included)
- The rest operation and maintenance cost of both SAEPs is in the range of 5.0 EURO per generated MWh.
- The land lease is not included in the calculation because it is a negligible cost for desert or semi desert installation

In order to calculate the FSC technology average direct production cost we can use the figures of the previous paragraph for the SAEP farm of 9 similar units. The dimensions of which are H=750m, d=40m and A_c=740000m². Each one of these SAEPs will have a rating power of 3MW and an annual generating ability of ~12.9GWh/year. Thus their construction cost was estimated to 6 million. The Annual repayment amount for each FSC SAEP will be 420000 EURO or a capital cost of 32.3 EURO per produced MWh/year.

For the concrete SAEP we consider as a moderate estimation the amount of 200 million EURO construction cost with an annual generation of ~100 GWh/year. Thus the annual repayment cost will be 15 million EURO or a capital cost of ~150EURO per MWh/year.

The fabric structure of the FSC should be replaced every 6÷10 years. Its replacement cost is estimated to be 50·H·d=1.5 million EURO (present value) or a maximum of 250000 EURO/year i.e. 19.2EURO MWh/year (for 6 year replacement period).

The rest operation and maintenance cost for both SAEPs is ~5 EURO per produced MWh. Thus the direct production cost of MWh/year by the two technologies is:

- FSC technology ~56.5 EURO/MWh
- Concrete solar chimney technology ~155 EURO/MWh

Both SAEP technologies operate 24 hours/day year round and they can replace the base load fossil fueled power plants (Coal, Natural Gas and Nuclear).

8.2 Direct production cost comparison

The following table (6) gives the comparison of the major electricity generating technologies. The figures for the rest technologies are average values of collected official data, released by EU authorities in various publications.

The conventional base load electricity generating technologies are the coal and the natural gas fueled technologies of combined cycle and the nuclear fission technology.The first two technologies are emitting greenhouse gases and should sooner or later be replaced by alternative zero emission technologies, while the third-one although it is of zero emission technology it is considered to be dangerous and health hazardous technology. A necessary condition for the replacement of the base load electricity generating technologies by alternative renewable technologies is that these alternative technologies should operate continuously and their sources should be unlimited. The nuclear fusion technology is an alternative but its progress is slow, while the global warming threat demands urgent actions. That goes too for the promising carbon capture and storage technology, besides the problems related to carbon dioxide safe sequestration

Fuel or Method of Electricity Generation	MWh Direct Production Cost in EURO	Investment in EURO per produced MWh/year	Mode of operation and Capacity factor
Coal fired (not including carbon emission penalties)	55-60	200	Combined cycle base load 85%
Coal fired with CCS (Carbon capture and storage)	80-100	300-400	Combined cycle base load 85%
Natural Gas fired (not including carbon emission penalties)	60-65	150	Combined cycle 85%
Nuclear Fission	65-75	400÷450	Base load 95%
Wind parks onshore	60	500	Intermittent 30%
Wind parks offshore	75	650	Intermittent 30%
Concentrating Solar CSP	180	2000	Continuous with thermal storage 30%
Photo Voltaic PV	280	3000	Intermittent 15-17%
Solar Chimney concrete	155	~2000	Continuous ~50%
Floating Solar Chimney	~60	~500	Continuous ~50%
Biomass	55-75	500-÷700	Continuous 85%
Geothermal	50-70	500-÷800	Continuous 90% (limited resource)
Hydroelectric	50-60	500÷800	Continuous (load following, limited resource)

Table 6. A cost comparison of electricity generating technologies

The wind and solar technologies are appropriate technologies if they are equipped with massive energy storage systems for continuous operation. With today's technology only the solar concentrating power plants (CSP) can be equipped with cost effective thermal energy storage systems and generate continuous electricity. However their MWh direct production cost is three times higher in comparison with the respective cost of the existing base load technologies. The FSC technology is by nature equipped with ground thermal storage and operates continuously. Due to its low investment cost and its almost equal direct production cost to the conventional base load electricity technologies it is an ideal candidate to replace the fossil fueled base load technologies.

9. Large scale application of the FSC technology in deserts

9.1 Desert solar technologies

The mid-latitude desert or semi desert areas of our planet are more than enough in order to cover the present and any future demand for solar electricity. According to most conservative estimations, a 3% of these areas with only 1% efficiency for solar electricity generation can supply 50% of our future electricity demand. Also these kinds of lands exist in all continents and near the major carbon emitting countries (USA, China, EU and India).

The desert solar technologies for continuous electricity generation are the following:
- The photo voltaic (PV) large scale farms equipped with batteries
- The concentrating solar power plants (CSP) equipped with thermal storage tanks
- The concrete solar chimney SAEPs or Solar Up-draft Towers
- The floating solar chimney (FSC) farms

The following table (7) is giving us a comprehensive comparison of these desert solar technologies (OM means operation and maintenance).

Desert Technology of continuous operation	Major benefits	Major problems	MWh Direct production cost in EURO	Investment per produced MWh/year
PV with energy storage batteries	-Demands no water -Low OM care and cost	-The replacement cost of the batteries	Very high 280	Very high >3000
CSP with thermal storage	-Low cost thermal storage	-Demands water for its operation -Demands OM personnel on site	High 180	High >2000
Solar up-draft Tower (concrete solar chimney)	- No water demand -High operating life -Low OM care and cost	-High initial cost -High construction period on site	High 155	High >2000
Floating Solar Chimney	-No water demand -Easy and fast deployment on site -Low OM care	-Periodic replacement of the FSC fabric parts	Low 60	Low 500

Table 7. Comparison of desert solar technologies

9.2 The Desertec project

The Desertec project is a proposal to EU for using the desert or semi desert areas in MENA area (Middle East and North Africa) in order to generate solar electricity. Using an appropriate area of 300KmX300Km in MENA with only 1% efficiency up to 50% of its present and future electricity demand can be generated.

The transmission of the generated electricity to the EU can be achieved by using UHVDC (Ultra High Voltage Direct Curent) lines. Using the existing technology up to 6.4 GW of electricity power can be transmitted by only one UHVDC line of two conductors (±800KV and 4000A).

The UHVDC lines can be overhead, underground or undersea lines with different construction costs but the same safety and reliability.

The farm of desert power plants generates AC electricity (up to 6.4 GW). This AC electricity is converted to DC electricity, at a special power station near the farm. Through a UHVDC line the DC electricity is transmitted to the chosen place of EU, where a reverse converter power station is transforming the DC to AC electricity with the suitable characteristics for the EU local grid.

The losses of the UHVDC transmission (including the losses of two converting power stations) are not more than ~5% per 2000 Km of transmission distance. Their construction cost for 2000Km average distance between MENA and EU areas, depends on the mode of the UHVDC line and will range between 1÷2 Billion EURO.

The following table shows a comparison cost for an electricity generation system of 6.4GW installed in MENA area and transmitting its electricity power to a EU grid for a distance of 2000Km. It is assumed that due to the energy storage systems of all the desert power plants their capacity factor is more or less similar (~50%). This practically means that the desert solar farms would generate electricity of ~6.4GW X (8760/2)hours≈28000GWh/year, of which ~95% or ~26500 GWh/year (or 26.5 TWh/year) will be transmitted to the EU chosen place.

In order to cover 40÷50% of the present and future EU electricity demand i.e. 1060÷1500 TWh/year we should build a set of 40 to 56 independent solar farms of 6.4GW that can be installed in appropriate MENA areas and connected through UHVDC lines to the proper places of EU countries. In order to build 40-56 farms we should invest capital of the amounts as shown in the next table (8) for respective technologies.

Desert Technology of continuous operation	Investment cost (including UHVDC lines cost of 1.5 billion EURO) for the solar farm of 6.4GW in billion EURO	Investment cost for building 40÷56 similar solar farms in billion EURO	MWh direct production cost in EURO (26.5 TWh supplied to EU)
PV with energy storage batteries	>85.5	3420 4778	>285
CSP (parabolic through or tower) with thermal storage	57.5	2300 3220	185
Solar up-draft Towers	57.5	2300 3220	160
Floating Solar Chimney	15.5	620 868	65

Table 8. Cost comparison of solar desert farms of 6.4 GW

The maximum desert or semi desert area for the installation of one solar farm of 6.4GW is not more than 1600 Km^2 or a square area ~(40Km X 40Km). Thus the maximum neaded area in order to cover the 40÷50% of the present and future EU electricity demand, with zero emission solar electricity, is 64000÷90000Km^2 (i.e. a square area of 250Km X 250Km up to 300Km X 300Km)

This maximum area is indispensable for solar chimney farms (concrete or floating) of 1% efficiency. As for the rest solar technologies a much smaller desert area is adequate. However the maximum area needed is not more than 2% of proper desert or semi desert area in MENA territory.

By the presented data it is evident that the FSC technology has tremendous benefits in comparison with its solar competitors for desert application.

Its major benefits are:

- Low investment cost
- Low KWh direct production cost (almost the same with the fuel consuming base load electricity generating technologies)
- 24hours/day uninterrupted operation due to the ground thermal storage
- The daily power profile can be as smooth as necessary using low cost additional thermal storage
- Demands no water for its operation and maintenance
- Easy and fast deployment on site
- It uses recycling and low energy production materials (mainly plastic and glass)
- Minimum personnel on site during its construction and operation

Large scale desert application of the Floating Solar Chimney technology can be one of the major tools for global warming elimination and sustainable development.

10. Climate change warning

Climate change indications due to the global warming threat are accelerating. Climate change policies should be agreed upon and urgent measures should be taken. Global warming due to greenhouse gases emissions (CO_2, CH_4 etc.) is a reality scientifically documented.

Intergovernmental Panel on Climate Change (IPCC) is a Nobel Priced UN committee studying carefully and objectively the global warming due to greenhouse gases produced by human activity on earth. The major producer is the fossil fuels used in residential, industrial, and transportation activities, of which the major-one is the electricity generation of fossil fueled power plants. According to IPCC estimations the global average temperature increase on earth will follow the pattern shown in the next figure (19) depending on our future model of energy use, electricity generation scenarios and greenhouse gases concentration. According to mentioned estimations, pertaining the existing technology and applying an internationally agreed upon strict policy on greenhouse gas emissions, the scenario most likely to come up is an eventuality between I and II.

According to mentioned scientifically documented estimations, global temperatures in excess of 1.9 to 4.6 ^0C warmer than pre-industrial would appear and it will be possibly sustained for centuries.

The major global warming effects on our planet, according to IPCC are:

- Anthropogenic warming and sea level rise would continue for centuries even if the greenhouse gas concentrations were to be stabilized
- Eventual melting of the Greenland ice sheet, would raise the sea level by 7 m compared to 125,000 years ago
- Due to precipitation changes fertile land devastation is possible to appear in many areas
- The existing atmospheric models can not exclude the appearance of extreme catastrophic atmospheric phenomena such as: very strong typhoons, tornados, snow or hail storms etc.

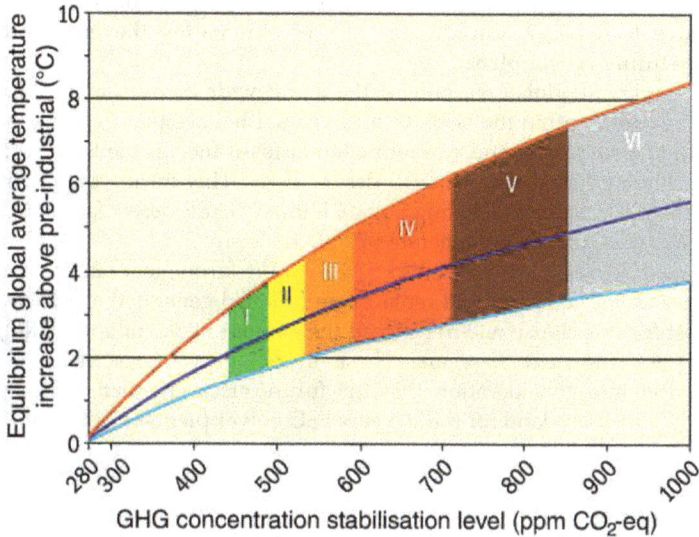

Fig. 19. IPCC scenarios of global temperature increase

The energy sector is the major source of the greenhouse gases due to its fossil fuelled technologies of electricity generation, transportation, industrial activities etc. For the year of 2010 an estimated quantity of 29,000 Mt of carbon dioxide will be spread all over the environment from fossil fuel combustion of which:

- 36.4 % from electricity generation
- 20.8 % from the industry
- 18.8 % from transport and
- 14.2 % from household, service and agriculture and
- 9.8 % from international bunkers

The mechanism of Kyoto protocol aims to create an "objective" over the external cost at least for the threatening carbon dioxide (CO_2) emissions through trading their rights.

The cost of the emitted CO_2, sooner or later it will reach at prices 20-30 EURO per ton of CO_2 and after the year 2012 for EU the fossil fuelled PPs should pay for each ton of CO_2 emitted by them. Taking into consideration that 1 Kg of coal has a thermal energy of ~8.14 KWh, thus a modern coal fired power plant with efficiency ~45% will generate by this ~ 3.66 KWh and will emit to the environment 3.667 Kg of CO_2. Thus in a modern coal fired plant approximately 1.0 Kg of CO_2 is emitted per generated KWh. For the lignite coal fired power plants this figure is 50% higher and for modern combined cycle natural gas power plants could be 50% smaller.

11. Conclusion

Although electricity generation is a major carbon dioxide producer we should notice that electricity can replace all the energy activities related to fossil fuelled technologies. Thus a solution to the global warming is possible if we succeed to generate zero emission clean electricity.

The renewable electricity generating technologies is a major tool, some believe that it should be the exclusive technology, towards the aim of eliminating the greenhouse emissions threatening the future on our planet.

It is possible to mitigate global warming if the world-wide consumption of fossil fuels can be drastically reduced within the next 10 to 15 years. I believe that the only viable scenario that could lead to a successful and real reduction of fossil fuels is the large scale application of the FSC technology in desert or semi desert areas. This means that we should start building, for the next 30 years, Floating Solar Chimney SAEP desert farms of overall rating power ~160 GW/year, that could generate ~720 TWh/year.

Thus for the next 30 years we will build SAEP desert farms generating more than 21600 TWh/year solar electricity that could replace fossil fuelled generated electricity. The global investment cost for this choice will not exceed the amount of 360 billion EURO/year or 11.5 trillion EURO for the next 30 years. These investments in electricity generation are reasonable taking into consideration that the future electricity demand could reach the 45000 TWh. The necessary land for the 30 years FSC power plants is 1.000.000 Km^2 (1000 Km X 1000 Km)

12. References

[1] Bernades M.A. dos S., Vob A., Weinrebe G., 2003 *"Thermal and technical analyses of solar chimneys"* Solar Energy 75 ELSEVIER, pp. 511-52.

[2] Backstrom T, Gannon A. 2000, *"Compressible Flow Through Solar Power Plant Chimneys"*. August vol 122/ pp.138-145.

[3] Gannon A. , Von Backstrom T 2000, *"Solar Chimney Cycle Analysis with System loss and solar Collector Performance"*, Journal of Solar Energy Engineering, August Vol 122/pp.133-137.

[4] Papageorgiou C. 2004 *"Solar Turbine Power Stations with Floating Solar Chimneys"*. IASTED proceedings of Power and Energy Systems, EuroPES 2004. Rhodes Greece, july 2004 pp,151-158

[5] Papageorgiou C. 2004, *"External Wind Effects on Floating Solar Chimney"* IASTED Proceedings of Power and Energy Systems, EuroPES 2004, Conference, Rhodes Greece ,July 2004 2004 pp.159-163

[6] Papageorgiou C. 2004, *"Efficiency of solar air turbine power stations with floating solar chimneys"* IASTED Proceedings of Power and Energy Systems Conference Florida, November 2004, pp. 127-134.

[7] Papageorgiou C. "Floating Solar Chimney" E.U. Patent 1618302 April. 29, 2009.

[8] Pretorius J.P., Kroger D.G. 2006,*"Solar Chimney Power Plant Performance"*, Journal of Solar Energy Engineering, August 2006, Vol 128 pp.302-311

[9] Pretorius J., *"Optimization and Control of a Large-scale Solar Chimney Power Plant"* Ph.D. dissertation, Dept. Mechanical Eng., Univ. Stellenbosch 7602 Matieland, South Africa 2007.

[10] Schlaich J. 1995, *"The Solar Chimney: Electricity from the sun"* Axel Mengers Edition, Stutgart

[11] J. Schlaich J. e.al 2005, *"Design of commercial Solar Updraft Tower Systems-Utilization of Solar Induced Convective Flows for Power Generation"* Journal of Solar Energy Engineering Feb. 2005 vol 127, pp. 117-124R.

[12] White F. *"Fluid Mechanics"* 4th Edition McGraw-Hill N.York 1999

2

New Trends in Designing Parabolic-Trough Solar Concentrators and Heat Storage Concrete Systems in Solar Power Plants

Valentina A. Salomoni[1], Carmelo E. Majorana[1], Giuseppe M. Giannuzzi[2],
Adio Miliozzi[2] and Daniele Nicolini[2]
[1]*University of Padua*
[2]*ENEA – Agency for New Technologies, Energy and Environment*
Italy

1. Introduction

Energy availability has always been an essential component of human civilization and the energetic consumption is directly linked to the produced wealth. In many depressed countries the level of solar radiation is considerably high and it could be the primary energy source under conditions that low cost, simple-to-be-used technologies are employed. Then, it is responsibility of the most advanced countries to develop new equipments to allow this progress for taking place. A large part of the energetic forecast, based on economic projection for the next decades, ensure us that fossil fuel supplies will be largely enough to cover the demand. The predicted and consistent increase in the energetic demand will be more and more covered by a larger use of fossil fuels, without great technology innovations. A series of worrying consequences are involved in the above scenario: important climatic changes are linked to strong CO_2 emissions; sustainable development is hindered by some problems linked to certainty of oil and natural gas supply; problems of global poverty are not solved but amplified by the unavoidable increase in fossil fuel prices caused by an increase in demand. These negative aspects can be avoided only if a really innovative and more acceptable technology will be available in the next decades at a suitable level to impress a substantial effect on the society. Solar energy is the ideal candidate to break this vicious circle between economic progress and consequent greenhouse effect. The low penetration on the market shown today by the existent renewable technologies, solar energy included, is explained by well-known reasons: the still high costs of the produced energy and the "discontinuity" of both solar and wind energies. These limitations must be removed in reasonable short times, with the support of innovative technologies, in view of such an urgent scenario.

On this purpose ENEA, on the basis of the Italian law n. 388/2000, has started an R&D program addressed to the development of CSP (*Concentrated Solar Power*) systems able to take advantage of solar energy as heat source at high temperature. One of the most relevant objectives of this research program (Rubbia, 2001) is the study of CSP systems operating in the field of medium temperatures (about 550°C), directed towards the development of a new and low-cost technology to concentrate the direct radiation and efficiently convert solar energy into high temperature heat; another aspect is focused on the production of hydrogen by means of thermo-chemical processes at temperatures above 800°C.

As well as cost reductions, the current innovative ENEA conception aims to introduce a set of innovations, concerning: i) *The parabolic-trough solar collector*: an innovative design to reduce production costs, installation and maintenance and to improve thermal efficiency is defined in collaboration with some Italian industries; ii) *The heat transfer fluid*: the synthetic hydrocarbon oil, which is flammable, expensive and unusable beyond 400°C, is substituted by a mixture of molten salts (sodium and potassium nitrate), widely used in the industrial field and chemically stable up to 600°C; iii) *The thermal storage* (TES): it allows for the storage of solar energy, which is then used when energy is not directly available from the sun (night and covered sky) (Pilkington, 2000). After some years of R&D activities, ENEA has built an experimental facility (defined within the Italian context as PCS, *"Prova Collettori Solari"*) at the Research Centre of Casaccia in Rome (ENEA, 2003), which incorporates the main proposed innovative elements (Figure 1). The next step is to test these innovations at full scale by means of a demonstration plant, as envisioned by the "Archimede" ENEA/ENEL Project in Sicily. Such a project is designed to upgrade the ENEL thermo-electrical combined-cycle power plant by about 5 MW, using solar thermal energy from concentrating parabolic-trough collectors.

Fig. 1. PCS tool solar collectors at ENEA Centre (Casaccia, Rome).

Particularly, the Chapter will focus on points i) and iii) above:

- loads, actions, and more generally, the whole design procedure for steel components of parabolic-trough solar concentrators will be considered in agreement with the Limit State method, as well as a new approach will be critically and carefully proposed to use this method in designing and testing "special structures" such as the one considered here;
- concrete tanks durability under prolonged thermal loads and temperature variations will be estimated by means of an upgraded F.E. coupled model for heat and mass transport (plus mechanical balance). The presence of a surrounding soil volume will be additionally accounted for to evaluate environmental risk scenarios.

Specific technological innovations will be considered, such as:

- higher structural safety related to the reduced settlements coming from the chosen shape of the tank (a below-grade cone shape storage);
- employment of HPC containment structures and foundations characterized by lower costs with respect to stainless steel structures;
- substitution of highly expensive corrugated steel liners with plane liners taking advantage of the geometric compensation of thermal dilations due to the conical shape of the tank;
- possibility of employing freezing passive systems for the concrete basement made of HPC, able to sustain temperature levels higher than those for OPC;
- fewer problems when the tank is located on low-strength soils.

2. Description of parabolic-trough solar concentrators

The parabolic-trough solar concentrators are one of the basic elements of a concentrating solar power plant. The functional thermodynamic process of a solar plant is shown in (Herrmann et al., 2004). The main elements of the plant are: the solar field, the storage system, the steam generator and the auxiliary systems for starting and controlling the plant. The solar field is the heart of the plant; the solar radiation replaces the fuel in conventional plants and the solar concentrators absorb and concentrate it. The field is made up of several collector elements composed in series to create the single collector line. The collected thermal energy is determined by the total number of collector elements which are characterized by a reflecting parabolic section (the concentrator), collecting and continuously concentrating the direct solar radiation by means of a sun-tracking control system to a linear receiver located on the focus of the parabolas. A circulating fluid flows inside a linear receiver to transport the absorbed heat.

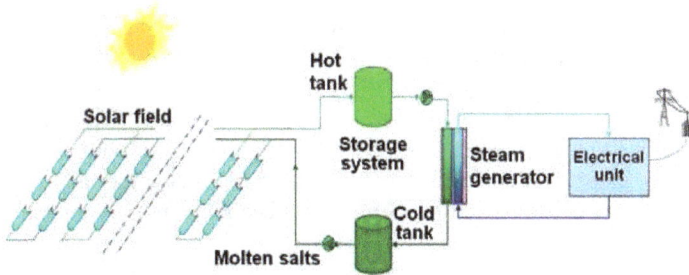

Fig. 2. Functional thermodynamic process flow of a solar plant.

A solar parabolic-trough collector line is divided into two parts from a central pylon supporting the hydraulic drive system (Antonaia et al., 2001). Each part is composed by an equal number of identical collector elements, connected mechanically in series. Each collector element consists of a support structure for the reflecting surfaces, the parabolic mirrors, the receiver line and the pylons connecting the whole system to a solid foundation by means of anchor bolts. The configuration of a solar parabolic-trough collector is that of a cylindrical-parabolic reflecting surface with a receiver tube co-axial with the focus-line, as a first approximation. The reflecting surface must be able to rotate around an axis parallel to the receiver tube, to constantly ensure that the incident radiation and the plane containing the parabolic sections' axes are parallel. In this way the incident solar light on the reflecting surfaces is concentrated and continuously intercepted by the receiver tube in any assumed position of the sun during its apparent motion. The parabolic-trough collector is then constituted by a rotating "mobile part" to orientate the concentrator reflecting surfaces and by a "fixed part" guaranteeing support and connection to the ground of the mobile part.

The solar collector performances, in terms both of mechanical strength and optical precision, are related to one side to the structural stiffness and on the other to the applied load level. The main load for a solar collector is that coming from the wind action on the structure and it is applied as a pressure distributed on the collector surfaces.

From a structural point of view, it must be emphasized that the parabolic-trough concentrator is composed mainly by three systems: the *concentration*, the *torque* and the *support system*. Other fundamental elements, not treated in this document for sake of brevity, are the *foundation* and the *motion systems*. In Table 1 the subsystems and basic elements characterizing the structure of the concentrator developed by ENEA are shown. All elements should be considered when designing a parabolic-trough concentrator and verified for "*operational*" and "*survival*" load conditions. Corrosion risks and safe-life (about 25-30 years) must be taken into account as well.

The following basic operational conditions, listed in Table 2, can be considered valid for a parabolic-trough concentrator; they define different performance levels under wind conditions. "*Design conditions*" can be fixed consequently.

Finally, on the basis of what described above, the main requirements when designing a parabolic-trough concentrator can be summarized as follows:

- *Safety*: the collector structures exposed to static loads must guarantee adequate safety levels to ensure public protection, according (in our case) to the Italian Law 1086/71. This is translated into a suitable strength level or more generally in safety factors for the construction within the *Limit State Analysis*.
- *Optical performance*: the structure must guarantee a suitable stiffness in order to obtain, under operational conditions, limited displacements and rotations, the optical performance level being related to the capacity of the mirrors concentrating the reflected radiation on the receiver tube.
- *Mechanical functionality*: the structural adaptation to loads must not produce interference among mobile and fixed parts of the structure under certain load conditions.
- *Low cost*: the structure has to respond to typical economic requirements for solar plant fields (e.g. known from experiences abroad): unlimited plant costs lead to non-competitive sources employments. This can lead to tolerate fixed damage levels of the structure under extreme conditions (i.e. collapse of not-bearing elements, local yield, etc.), but still respecting the above mentioned requirements of public protection.

3. Codes of practice and rules

The parabolic-trough concentrator, on the basis of its structural shape and use and further considering available National and European recommendations, is classifiable as a "*special structure*" (Majorana & Salomoni, 2004 (a); Giannuzzi et al., 2007): it is not a machine or a standard construction. The definition "special" comes directly from a subdivision in classes and categories according to the criterion of the "Rates for professional services" as it results from the Italian law n. 143/1949; this law places "Metallic structures of special type, notable constructive importance and requiring ad-hoc calculations" into class IX e subclass b.

Systems	Subsystems	Elements
Concentration system	Reflecting surfaces	Mirrors, Mirror–structure connection
	Mirrors support structures	Girders, Girder-framed structure connection
		Framed structure, Framed structure–torque tube connection
Torque system	Torque tube, plate, hinge	Torque tube, Torque tube–plate connection, Plate, Plate–hinge connection, Hinge
Module supports	Intermediate / final pylons	Cylindrical pin joint, Pin joint–support connection, Framed structure, Plate, Anchor bolts
	Central pylon	Cylindrical pin joint, Pin joint–support connection, Framed structure, Engine support structure, Plate, Anchor bolts
Other	Foundations	Piles and/or plinths, Anchor bolts
	Drive system	Hydraulic drive/pistons, etc.

Table 1. Example of structural elements of a parabolic-trough concentrator.

Level	Condition
W1	Response under normal operational conditions with light winds. The concentration efficiency must be as high as possible under wind velocity less than a value v_1 characterizing this level.
W2	Response under normal operational conditions with medium winds. The concentration efficiency is gradually diminishing under wind velocity comprised between v_1 and v_2. The wind velocity v_2 characterizes this level.
W3	Transition between normal operating conditions and survival positions under medium-to-strong or strong winds. The survival must be ensured in any position under medium–strong winds. The drive must be able to take the collector to safe positions for any wind velocity comprised between v_2 and v_3. The wind velocity v_3 characterizes this level.
W4	Survival under strong winds in "rest" positions. The survival wind velocity must be adapted to the requests of the site according to recommendations. The wind velocity v_4 characterizes this level.

Table 2. Operational conditions.

From the functional analysis of the structure its special typology clearly emerges, according to its design, technical arrangements and innovation. When the parabolas are stopped in an assigned angular configuration, the nature of the structure can be determined: steel structure of mixed type founded on simple or reinforced concrete placed on a foundation soil having characteristics closely correlated to a chosen site, also under the seismic profile. From the structural point of view, the dynamic characteristics play a major role, with the response deeply influenced not only by the drive-induced oscillations, but also by dominant winds or seismic actions. Taking into account the above considerations, it is then possible to state that the examined structure is "*special*".

Moreover, such a structure requires appropriate calculations since some parts are mobile, even if with a slow rotation; at the same time the structure is subjected to wind actions, especially relevant due to the parabolas dimension. The simultaneous thermal and seismic actions, acting as self-equilibrated stresses in an externally hyperstatic structure, are equally important. Special steel made structures are e.g. cranes: they are designed using specific recommendations; in our case the reference to existing codes of practice is necessary, even if with the aim of adapting them and/or proposing new ones for CSP systems. Hence it clearly appears that such structures, built within the European countries, are currently designed and verified out of standards; the only two Italian recommendations acting as guidelines are:

- Law 5/11/71, n.1086, Norms to discipline the structures made by plain and pre-stressed reinforced concrete and by metallic materials.
- Law 2/2/74, n.64, Procedures devoted to structures with special prescriptions for seismic zones.

Moreover, several "technical norms" are related to the above ones, in form of "Minister (of Public Works) Decrees", or "explanation documents", or other documents giving rise to a certain amount of duplications and repetitions; however, a progressive compulsory use of Eurocodes is being introduced to push Italian engineers more properly into the European environment. In this case, Eurocodes 3 and 8 are of interest for the structural design of solar concentrators, also in view of their seismic performance. It is important to make an advanced choice regarding the body of recommendations to be followed in the design and checking phases and to proceed further with them, avoiding the common mistake of some designers to take parts from one norm (i.e. Italian) and mix it with parts of another norm (i.e. Eurocodes). The main problems in the so-called harmonization of rules within Europe reside in finding safety coefficients to be applied for considering special conditions (e.g. environmental) in each country, as well as those for materials. This is a source of difficulty in the creation of a unique body of rules valid in the whole European territory. The last product of recommendations recently emitted by the actual Ministry of Public Works in Italy is a 438 pages document (plus two Annexes) named "*Testo Unico per le Costruzioni*". It is compulsory in the Italian territory from July 1st 2009. The aim of this decree was also to unify a series of previous decrees into a single document. As already stated, it has been here chosen to follow the current Italian laws, and Eurocodes for comparison, in view of the possible application of solar concentrators at Priolo Gargallo (near Syracuse, Sicily). In principle, with a few changes, it is possible to apply the technology in other sites, as well as outside Italy or even Europe: slight changes in dimensioning could occur.

Hence, to take into account the specificity of the investigated structures, it was necessary to combine together *operational states* (OSs) (Table 2), *characteristic positions* and *load actions*, reaching to the interpretation of Table 3 within the context of a *limit state* (LS) analysis (Salomoni et al., 2006). Additionally, within the *serviceability limit states* (SLSs) the conditions of maximum rotation (W_1 operational state) and maximum deformation (W_2) must be verified; W_3 requires the collector operability within an elastic *ultimate limit state* (ULS), i.e. absence of permanent deformations. Differently, such deformations can be present within W_4 but without leading to a structural collapse.

Characteristic positions		→						
Operational states			-120°	-30°	0°	30°	60°	75°
V_{ref} [m/sec]	↓	Limit state						
7	W_1	SLS		Y				
14	W_2	SLS					Y	
21	W_3	ULS	Y	Y	Y	Y	Y	Y
28	W_4	ULS	Y		Y			

Table 3. Example of combinations among characteristic positions, operational states and load actions to study CSPs in the context of LS analyses.

4. Materials

The solar concentrator supporting structure is made of hot-laminated steel. Hence, according to Eurocode 3 and UNI EN 10025, steels in form of bars, plates or tubes must be of the types shown in Table 4.

However recommendations allow for using different types of steel once the ensured safety level remains the same, justifying this through appropriate theoretical and experimental documentations. Under uniaxial stress states, their design strengths can be deduced from tables; in case of multiaxial states, suitable combinations are additionally given. In our calculations, the following material characteristics are considered: elastic modulus E = 210000 N/mm², Poisson's coefficient v = 0.3, thermal expansion coefficient α = 12•10-6 °C-1 and density ρ = 7850 kg/m³. If welding is used for connecting elements, the behaviour of steel types S235 and S275 is distinguished from that of S360.

Nominal steel type	thickness t [mm]			
	t ≤ 40		40 < t ≤ 100	
	f_y [N/mm²]	f_u [N/mm²]	f_y [N/mm²]	f_u [N/mm²]
Fe360 / S235 (EN 10025)	235	360	215	340
Fe430 / S275 (EN 10025)	275	430	255	410
Fe510 / S360 (EN 10025)	360	510	335	490

Table 4. Strengths and failure stresses (nominal values) for structural steels.

5. Loads

Given the design loads, subdivided in *permanent* and *variable* ones, wind conditions are here examined more in detail, whose effects on the structure are connected to the parabolas aerodynamics in their different characteristic positions (see below). The role of the snow has been additionally considered.

5.1 Variable loads
5.1.1 Wind action on the parabolas
The mean value of wind velocity, as a function of the distance from soil $V_m(z)$, is expressed by

$$V_m(z) = C_r(z) \cdot C_t(z) \cdot V_{ref} \tag{1}$$

where V_{ref} is the reference wind velocity, $C_r(z)$ the roughness coefficient and $C_t(z)$ the topographic coefficient.

The reference wind velocity V_{ref} is defined as the mean wind speed over a time period of 10 min, at 10 m height on a second category soil, with a 50 years "return period". The reference wind speeds for each Italian area is given by recommendations; e.g. a site located near the sea in Southern Italy has a reference wind speed of about 28 m/s. An important wind speed value is the peak wind speed which can be seen as the superposition of the mean wind speed plus its variation due to turbulence conditions on site. It can be evaluated as

$$V_{peak}(z) = G(z) \cdot V_m(z) \tag{2}$$

where $G(z)$ is the "peak factor", that is,

$$G = \sqrt{1 + \frac{7}{C_t(z) \cdot \ln(z/z_0)}} \tag{3}$$

Usually G is comprised between 1.5 and 1.6. It should be emphasized that the check under failure loads must be necessarily performed on the basis of the peak velocity, since this gives an overload capable of making the material reach its strength limit, even if its duration is short. As far as the operational performance is concerned, it is more feasible to use the mean velocity. The roughness coefficient $C_r(z)$ takes into account the variability of the mean wind speed and the site characteristics by considering the height over the soil and the soil roughness as functions of the wind direction. The roughness coefficient at height z is defined by the logarithmic profile

$$C_r(z) = k_r \ln(z/z_0) \tag{4}$$

where k_r is the soil factor and z_0 is the roughness length, both related to the soil exposure category on its turn linked to the geographic location of the investigated area within Italy and on the basis of the soil roughness. In case of an open country, k_r is 0.19 and z_0 is 0.05 m. The topographic coefficient $C_t(z)$ takes into account the increment in the mean wind speed on escarpments and isolated hills; in our case $C_t = 1$ can be taken.

The solar concentrator shape is taken into account by means of aerodynamic coefficients. The different aerodynamic shape coefficients have been identified by means of a CFD analysis carried out in (Miliozzi et al., 2007). These coefficients have been determined starting from wind actions exerted on the linear parabolic collector as functions of its angular position (Figure 3). Such coefficients have been calculated for the most (external) and the least (internal) stressed collectors (Giannuzzi, 2007), see e.g. Figure 4. An external collector is one of those belonging to the first line without any artificial barrier against wind actions, whereas an internal collector is one on the sixth line, taken as representative of all the others.

Full tables for shape coefficients in case of "external" parabolas as well as "internal" ones are reported in (Majorana & Salomoni, 2005 (a)) and used in (Majorana & Salomoni, 2005 (b)) for structural assessment within the Limit State Design. Shape coefficients have been used to evaluate drag (C_{fx}), lift (C_{fy}), torsion (C_{Mz}) and mean pressure (C_{pm}), each of them being function of the concentrator rotation angle, where the allowed rotation is in the range +/- 120°. Then, shape coefficients for mean pressures have been calculated as functions of the aperture angle for "external" or "internal" parabolas. By analyzing the above coefficients it is possible to identify the parabolas' characteristic positions listed in Table 5.

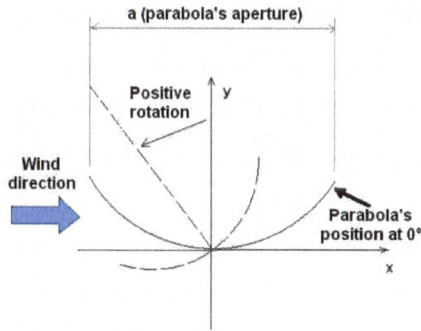

Fig. 3. Parabolic concentrator scheme at different angular positions.

Starting from the calculated shape coefficients, the corresponding effects referring to drag and lift force, torsion, mean pressure and pressure distribution have been determined.

By analyzing the results of the CFD analysis, it has been evidenced that aerodynamic coefficients and associated loads are largely reduced at the internal collectors. The main reason resides in the shielding effect produced by the first collectors' rows. This remark leads to the necessity of designing "strong" collectors along the external rows (Figure 4) and "light" collectors along the internal ones. Alternatively, it is possible to choose a different

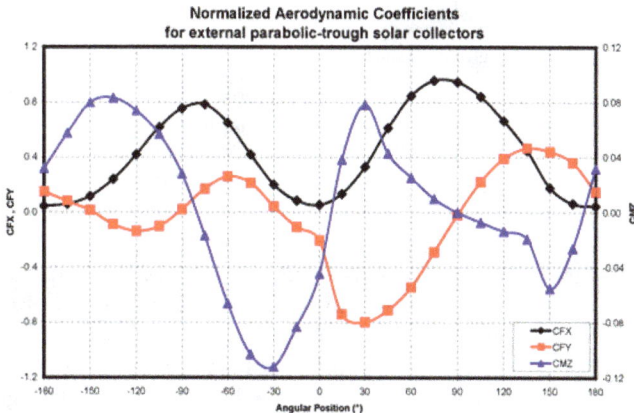

Fig. 4. Angular distribution of the normalized shape coefficients for "external" parabolas.

design strategy, based on the introduction of opportune windbreak barriers and on the realization of "light" collectors only. The position characterized by smaller loads is at 180°. This is only a theoretical, unattainable position because of the interferences between receivers and pylons. The safety position to be really taken in consideration is at about -120°. The waiting position (at 0°) does not guarantee an adequate level of protection for the mirrors. All the positions shown in Table 5 must be taken into account during the design phase but the most relevant position is, without doubt, the one associated to the maximum torque action. This is consequence of the fact that torque effects are accumulated along all the line, producing the maximum stresses on the structural elements close to the central pylon. This can be considered the key action in the parabolic-trough solar concentrators wind design.

5.1.2 Snow
The snow load is usually evaluated on the roofs (here parabolas), by means of the following expression

$$q_S = \mu_i \cdot q_{Sk} \tag{5}$$

where q_s is the snow load on the roof, μ_i the roof shape coefficient and q_{sk} the reference value of the snow load on the ground.

Characteristic Effect	Angular position (degrees)	
	"External" collector	"Internal" collector
Safety position	-120	-120
Waiting position	0	0
Maximum torque effect	-30	-15
Maximum bending action on the torque tube	+60	+30
Maximum drag force	+75	-45
Maximum lift force	+120	-45
Maximum crush force	+30	+30

Table 5. Wind effect: characteristic positions.

The load acts along the vertical direction and it is referred to the horizontal projection of the covering surface. The snow load on the ground depends on local environmental and exposure conditions, where the variability of the snowfall from region to region is taken into account. The reference snow load in locations at heights less than 1500 m over the mean sea level (m.s.l.) has to be evaluated on the basis of given expressions (whose values correspond to a "return period" of about 200 years). In case of a region like Sicily and a site located at a reference height less then 200 m m.s.l., q_{sk} is about 0.75 kN/m². The shape coefficients to be used for the snow load are those indicated in Table 6, being α (degrees) the angle between cover and the horizontal plane.

The shape coefficients μ_1, μ_2, μ_3, μ_{1*} refer to roofs having one or more slopes, and they should be evaluated as functions of α, as indicated by the codes. For given parabolas positions, other coefficients can be used, as e.g. those related to cylindrical covers. In absence of rifting inhibiting snow sliding, for cylindrical covers of any shape and single curvature of constant sign, the worst uniform and not-symmetric load distribution is there considered.

	$0° <= α <= 15°$	$15° < α <= 30°$	$30° < α <= 60°$	$α > 60°$
$μ_1$	0.8	0.8	$0.8(60-α)/30$	0.0
$μ_2$	0.8	$0.8+0.4(α-15)/30$	$(60-α)/30$	0.0
$μ_3$	$0.8+0.8α/30$	$0.8+0.8α/30$	1.6	-
$μ_1^*$	0.8	$0.8(60-α)/45$		0

Table 6. Shape coefficient for the snow load (Eurocode1-Part 2.3).

In our case, to determine the shape coefficients $μ_i$ for the parabolas, it is possible to approximatively evaluate the maximum slope of the parabolic collector with respect to the horizontal line, if it is rotated with its concavity upwards, being the element profile defined by means of the equation

$$y = x^2 / 4f \tag{6}$$

where $-2950 < x < 2950$ (mm), $f = 1810$ (mm); and the slope by

$$y' = x / 2f \tag{7}$$

with maximum value equal to 0.815, corresponding to an angle $α$ such that $tgα = 0.815$, i.e. $α ≈ 39°$. On the other side, taking into account the value corresponding to $x/2$, then $α = 22°$. Hence, assuming $α = 22°$ as a mean value, it is possible to calculate the shape coefficients as indicated in the recommendations. The corresponding load conditions are shown in Figure 5.

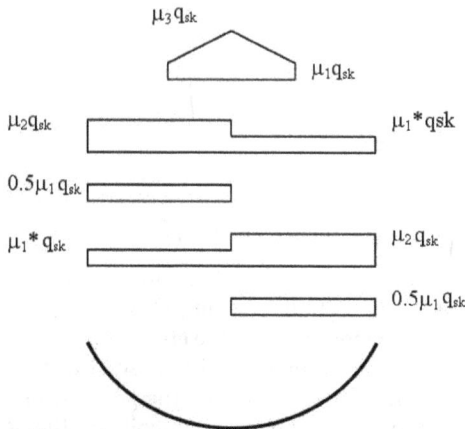

Fig. 5. Snow conditions for the parabolas when the solar collector is rotated to the waiting position.

As demonstrated in (Majorana & Salomoni, 2005 (b)), snow effects are fundamental when verifying the structure in the safety position (Table 5) or when seismic effects are included.

Discussing about the real significance of such an effect when considering desert locations (as those typical for CSP systems) is reasonable: this should be another example of the necessity for ad-hoc codes of practice when studying *special* structures in possibly *special* sites.

The parabola's configuration with its concavity upwards (Figure 5) is not the only possible one when evaluating the effects of the snow; being the snowy phenomenon largely predictable, so that a rotation of the collector towards the safety position is expected, an additional investigated angular position for analyzing snow effects refers to $\alpha = \pm120°$. When e.g. $\alpha = +120°$, the situation is the one of Figure 6; the remaining characteristic positions, even associable to different OSs, can be considered as characterized by a null snow action: in fact, in case of snow, the collector would be evidently moved to its safety position with no tracking. Additionally, being L_1 (distance between the point, on the rotated parabola, with null tangent and the origin of the vertical axis) > L (= 2950 mm in our case), it is precautionarily assumed $L_1 = L$ and hence from Eurocodes $\mu_1 = 0.8$, $\mu_2 = 2.0$ ($\mu_3 = 1.0$), with loads as those of Figure 7.

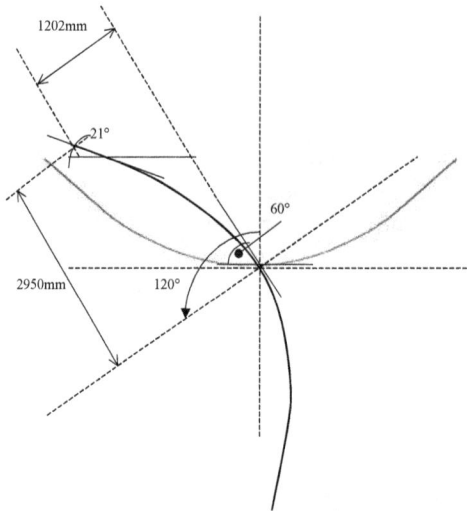

Fig. 6. Parabola's position for evaluating snow effects.

Two changes have been essentially introduced to what indicated by the codes: being, as already stated, $L_1 > L$, the point of null load amplified by μ_2 goes outside the effective parabola's projected dimension (consequently, the effect of μ_3 is zero; anyway, we are still in favour of safety being $\mu_2 > \mu_3$) and this explains the chosen trapezoidal shape for the load of Figure 7; the load cusp (from Eurocodes falling on the point whose slope on the curve is of 30°, i.e. at $L_1/4$), considered the not-symmetric parabolic profile, is moved with respect to $L_1/4$.

Then, when combining the loads, among the various indicated load conditions only those revealed as heaviest for the structural system have been adopted.

Hence, the main load combinations are reported in Table 7, where the multiplicative coefficients related to each basic action (permanent, G_k, and variable, Q_k) and to strength (f_y) are additionally indicated, for the OSs of Table 3.

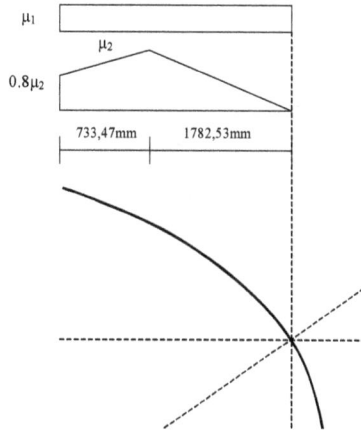

Fig. 7. Snow condition for the parabola when the solar collector is rotated to the safety position.

Combinations		G_k	Q_{1k} (wind)	Q_{2k} (snow)	f_y
W_130	W_130R	1.	1.	0.	1.
W_260	W_260R	1.	1.	0.	1.
$W_3\alpha$ $-120° \leq \alpha \leq 75°$ (Table 3)	$W_3\alpha E$	1.4*	1.5	0.	1.
$W_4\beta ; \beta = 0°$ (Table 3)	$W_4\beta E$	1.4*	1.5	0.	1.
$W_4\beta ; \beta = -120°$ (Table 3)	$W_4\beta E$	1.4*	1.5	0.	1.
		1.4*	1.5	1.05	1.
		1.4*	1.05	1.5	1.
		1.4*	0.	1.5	1.
	$W_4\beta P$	1.4*	1.5	1.05	0.83
		1.4*	1.05	1.5	0.83

Table 7. Main load combinations and corresponding multiplicative coefficients (*: if not acting in favour of safety; R: rare; E: elastic limit state; P: plastic collapse).

Particularly, for combinations related to states W_130 (W_1, 30°) and W_260, just *rare* ones are considered, being *frequent* and *quasi-permanent* combinations already included.
In the following, the main results related only to the concentration system are reported, being the conducted design and analysis methodology repeatable in the same way to the other macro-systems, i.e. the torque system and the module's support.

6. Analysis and verification of the concentration system

The concentration system is composed by three main elements: centering, stringers and reflecting mirrors (Figures 8 and 9). The system has been analysed considering a single modulus of 12 m, reproducing also the torque tube to which the centerings are linked.

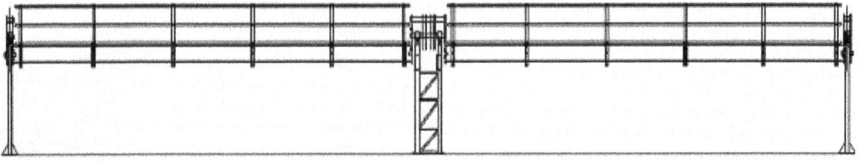

Fig. 8. Sketch of the solar collector (portion).

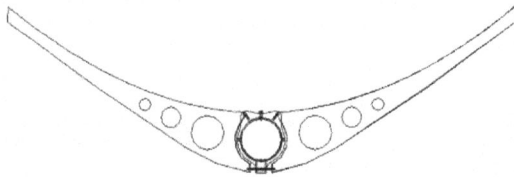

Fig. 9. Sketch of a typical centering (first proposal).

6.1 Limit states and load combinations

As already reported in the previous Section, the considered limit states and load combinations are summarized in Table 8. Correspondingly, OSs W_1 and W_2 are associated to SLSs for which the wind loads refer to a medium velocity and the serviceability limits referring to maximum torsion and maximum deformation, respectively, must be verified. Differently, in the ULSs W_3 and W_4 the structural permanence within the elastic state as well as tightness under loads corresponding to a characteristic peak wind must be verified. Particularly, in the W4 state the possible presence of snow has to be additionally accounted for. For both ULSs, a structural instability verification has to be conducted.

Operational states	V_{ref} (m/s) @10m	Limit state	Reference velocity
W_1	7	Serviceability	Medium
W_2	14	Serviceability	Medium
W_3	21	Ultimate, Elastic	Peak
W_4	28	Ultimate, Collapse	Peak

Table 8. Summary of adopted limit states and load combinations for the concentration system.

In Table 9 all the possible load combinations are shown which have been considered for developing the above-mentioned verifications. It is to be noticed that, for the collapse ULS, in addition to the combinations required by the Recommendations, two other combinations have been evaluated in which the snow and the wind alone are present: this was necessarily done due to the fact that the concurrent presence of the two loads, if from one side it increases the acting forces, from the other it reduces the magnitude of the torque bending, so reducing the stress state in some fundamental structural components. All the analyses have been performed in an elastic state and just in those cases, corresponding to a collapse ULS, in which the structure is particularly stressed, a tightness evaluation within a plastic state has been conducted.

LS	OS	Angle (°)	G_receiver	G_conc	Q_wind	Q_snow	c_G	c_Qv	c_Qn	f_y	ID
ELS	W_1	30	yes	yes	yes	no	1.00	1.00	0.00	1.00	w1p030c1
ELS	W_2	60	yes	yes	yes	no	1.00	1.00	0.00	1.00	w2p060c1
ULS elastic	W_3	75	yes	yes	yes	no	1.40	1.50	0.00	1.00	w3p075c1
							1.00	1.50	0.00	1.00	w3p075c2
		60	yes	yes	yes	no	1.40	1.50	0.00	1.00	w3p060c1
							1.00	1.50	0.00	1.00	w3p060c2
		30	yes	yes	yes	no	1.40	1.50	0.00	1.00	w3p030c1
							1.00	1.50	0.00	1.00	w3p030c2
		0	yes	yes	yes	no	1.40	1.50	0.00	1.00	w3p000c1
							1.00	1.50	0.00	1.00	w3p000c2
		-30	yes	yes	yes	no	1.40	1.50	0.00	1.00	w3m030c1
							1.00	1.50	0.00	1.00	w3m030c2
		-120	yes	yes	yes	no	1.40	1.50	0.00	1.00	w3m120c1
							1.00	1.50	0.00	1.00	w3m120c2
ULS collapse	W_4	-120	yes	yes	yes	load1	1.40	1.50	1.05	0.83	w4m120c1
							1.00	1.50	1.05	0.83	w4m120c2
							1.40	1.05	1.50	0.83	w4m120c3
							1.00	1.05	1.50	0.83	w4m120c4
							1.40	1.50	0.00	0.83	w4m120c5
							1.00	1.50	0.00	0.83	w4m120c6
							1.40	0.00	1.50	0.83	w4m120c7
							1.00	0.00	1.50	0.83	w4m120c8
		-120	yes	yes	yes	load2	1.40	1.50	1.05	0.83	w4m120c9
							1.00	1.50	1.05	0.83	w4m120c10
							1.40	1.05	1.50	0.83	w4m120c11
							1.00	1.05	1.50	0.83	w4m120c12
							1.40	1.50	0.00	0.83	w4m120c5
							1.00	1.50	0.00	0.83	w4m120c6
							1.40	0.00	1.50	0.83	w4m120c13
							1.00	0.00	1.50	0.83	w4m120c14
		0	yes	yes	yes	no	1.40	1.50	0.00	0.83	w4p000c1
							1.00	1.50	0.00	0.83	w4p000c2

Table 9. Details of the load combinations for the concentration system.

6.2 Analysis methodologies.

The structural element has been studied through the F.E. Cast3M code, realizing a 3D model of the 12 m concentration system (Figure 10). Reflecting mirrors, centerings, stringers, torque tube and edge plates. Apart from the plates, which have been modelled through infinitely-rigid beams, all the other components, being made by thin plates, have been modelled through 2D shell elements, able to take into account membrane as well as bending and shear stresses.

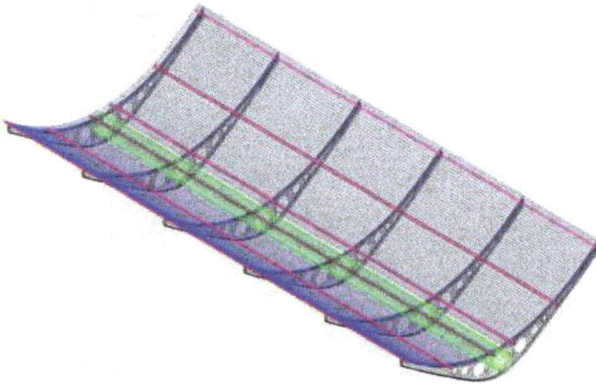

Fig. 10. 3D F.E. model of the concentration system.

The global structural constraints, applied to the edges of the connecting plates, have been applied such to create an isolated and isostatic system, so the stress state doesn't change due to possible loads transmitted by the adjacent moduli.

6.3 Discussion of the main numerical results.

The main results referring to SLSs for weak and medium wind, as well as to ULSs (elastic and collapse) are depicted in Table 10; stresses are calculated as the maximum equivalent Tresca stress, F_{saf} is the safety factor obtained by dividing the material yield limit (reduced in case of ULS, see Tables 7 and 9) by the above stress. The medium value of the parabola's deformation is additionally reported (which is always lower than ± 5 mm, the assumed limit within SLSs).

It is hence evidenced that in both elastic and collapse ULSs the safety factors are generally lower than one; by examining the results in more detail, it has been found that local yielding occur in the higher and middle part of the centering and in some zones connecting the centering with the stringers.

Anyway, it is to be said that the model has been developed to study the global stress level in the various components and not to locally analyse the connecting constructive details which need specific 3D models; a possible local overcome in the stress yield limit and/or consequent re-distributions of stresses can't be caught by such an approach, as explained below.

For sake of brevity, the contour maps of stresses have been included (Figures 11 and 12) referring only to w3p060c1 and w4m120c9 combinations: it is here evidenced the much localized nature of plasticization, as explained above.

	Component	σ_{max}	F_{sat}	Δ (mm)
w1p030c1	Stringers	87	3,17	2,33
	Centerings	108	2,54	
w2p030c0	Stringers	120	2,29	3,66
	Centerings	148	1,86	
w2p060c1	Stringers	117	2,36	3,05
	Centerings	142	1,94	
w3p075c1	Stringers	545	0,51	
	Centerings	597	0,46	
w3p075c2	Stringers	542	0,51	
	Centerings	578	0,48	
w3p060c1	Stringers	571	0,48	
	Centerings	622	0,44	
w3p060c2	Stringers	563	0,49	
	Centerings	602	0,46	
w3p030c1	Stringers	549	0,50	
	Centerings	604	0,46	
w3p030c2	Stringers	534	0,51	
	Centerings	566	0,49	
w3p000c1	Stringers	239	1,15	
	Centerings	262	1,05	
w3p000c2	Stringers	214	1,29	
	Centerings	232	1,19	
w3m030c1	Stringers	218	1,26	
	Centerings	221	1,24	
w3m030c2	Stringers	217	1,27	
	Centerings	183	1,50	
w3m120c1	Stringers	316	0,87	
	Centerings	299	0,92	
w3m120c2	Stringers	305	0,90	
	Centerings	285	0,97	
w4m120c1	Stringers	580	0,47	
	Centerings	564	0,49	
w4m120c2	Stringers	570	0,48	
	Centerings	540	0,51	
w4m120c3	Stringers	453	0,61	
	Centerings	494	0,56	

	Component	σ_{max}	F_{sat}
w4m120c1	Stringers	580	0,47
	Centerings	564	0,49
w4m120c2	Stringers	570	0,48
	Centerings	540	0,51
w4m120c3	Stringers	453	0,61
	Centerings	494	0,56
w4m120c4	Stringers	442	0,62
	Centerings	470	0,59
w4m120c5	Stringers	532	0,52
	Centerings	500	0,55
w4m120c6	Stringers	522	0,53
	Centerings	505	0,54
w4m120c7	Stringers	237	1,16
	Centerings	333	0,83
w4m120c8	Stringers	210	1,31
	Centerings	298	0,92
w4m120c9	Stringers	615	0,45
	Centerings	639	0,43
w4m120c10	Stringers	605	0,45
	Centerings	615	0,45
w4m120c11	Stringers	531	0,52
	Centerings	612	0,45
w4m120c12	Stringers	524	0,53
	Centerings	577	0,48
w4m120c13	Stringers	392	0,70
	Centerings	490	0,56
w4m120c14	Stringers	386	0,71
	Centerings	454	0,61
w4p000c1	Stringers	357	0,77
	Centerings	383	0,72
w4p000c2	Stringers	332	0,83
	Centerings	353	0,78

Table 10. Numerical results (static analyses) for the concentration system.

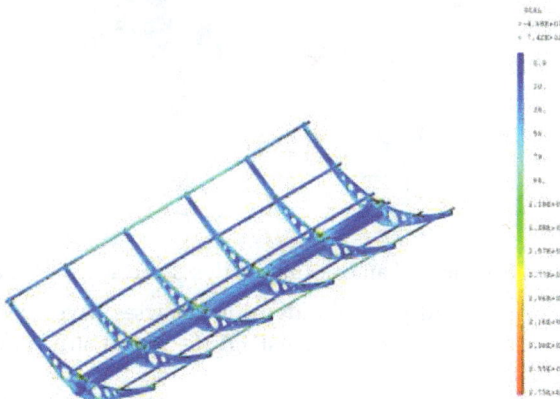

Fig. 11. Contour map of maximum equivalent Tresca stresses for w3p060c1.

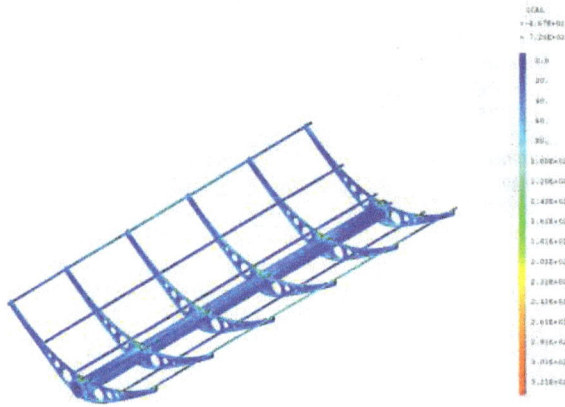

Fig. 12. Contour map of maximum equivalent Tresca stresses for w4m120c9.

The 3D static analyses revealed an appropriate response of the structure under a variety of actions and once, for example, the material strength had been locally overcome, appropriate design procedures have been updated and nonlinear (for material and geometry) analyses performed (see e.g. Figure 13).

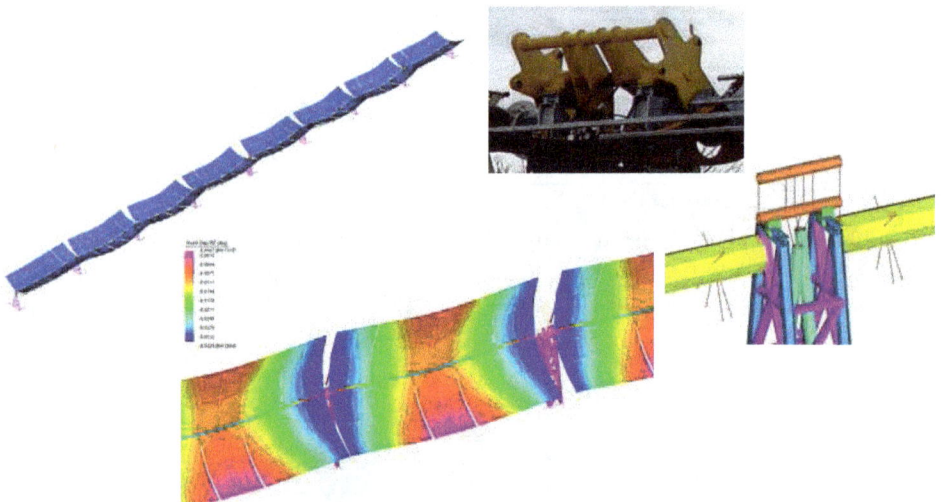

Fig. 13. Typical results from modal and seismic analyses and scheme for a modelled joint.

Additional modal, spectral and generally dynamic analyses have been conducted for the whole CSP system (see Figure 14) to understand the global structural behaviour and to newly upgrade the first design sketch.

The discussion about such results and the corresponding structural response can't be reported here for sake of brevity.

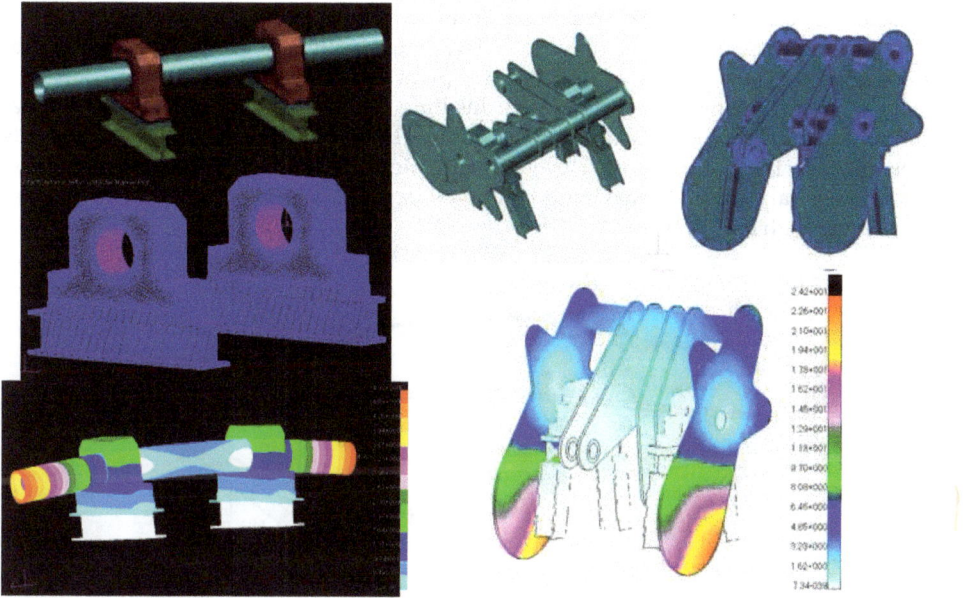

Fig. 14. Joints, pins and specific nodes studied through 3D nonlinear analyses for material and geometry to test their effective structural response and to verify the requirements of the different operational states.

7. Description of heat storage concrete systems

The main advantage of thermal solar power plants is the possibility to use relatively economical storage systems, if compared to other renewable energies (i.e. photo-voltaic and wind). Storing electricity is much more expensive than storing thermal energy itself. Thermal Energy Storage (TES) option can collect energy in order to shift its use to later times, or to smooth out the plant output during irregularly cloudy weather conditions. Hence, the functional operativeness of a solar thermal power plant can be extended beyond periods of no solar radiation without the need of burning fossil fuel. Periods of mismatch among energy supplied by the sun and energy demand can be reduced. Economic thermal storage is a technological key issue for the future success of solar thermal technologies.

In our days, among eight thermal storage systems in thermo-electric solar plants, seven have been of experimental or prototypal nature and only one has been a commercial unit (Salomoni et al., 2008). All the considered systems are "at sensible heat storage": two single-tanks oil thermo-cline systems, four two-tanks single medium systems (one oil- and three molten salt-) and two single-tanks double medium systems. Actually the most advanced technology for heat storage in solar towers and through collector plants considers the use of a two-tanks molten salt system (Ives et al., 1985).

Generally, the hot and cold tanks are located on the ground and they are characterized by an internal circumferential and longitudinally-wrinkled liner, appropriately thermally insulated. The cost of the liner is the primary cost of such a tank. In recent studies it has been shown that an increase in the hourly capacity accumulation reduces sensibly the cost of the produced electrical energy (LEC); this leads to increase the reservoir dimensions from the 11.6 m diameter and 8.5 m height of the Solar Two power plant to the larger 18.9 m diameter and 2.5 height calculated in the Solar Tres power plant design phase.

Already in 1985, the Solar Energy Research Institute (SERI) commissioned the conceptual design of a below-grade cone shape storage (Figure 15) with 900°C molten carbonate salts (Copeland et al., 1984). This solution, even though interesting because of the use of low cost structural materials, showed some limits connected to the high level of corrosion induced by carbonate and high temperature.

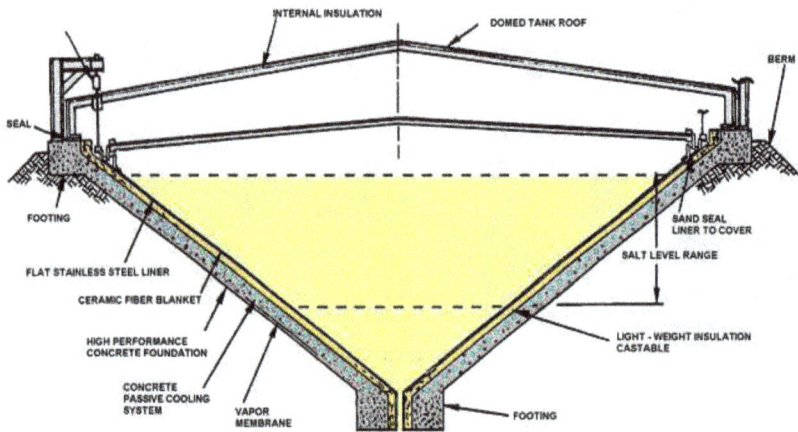

Fig. 15. Conical storage partially buried in the ground.

Such a type of storage is here reconsidered in combination with nitrate molten salts at a maximum temperature of 565°C, using an innovative high performance concrete (HPC) for the tanks. From the technological point of view, the innovations rely in:

- higher structural safety related to the reduced settlements;
- employment of HPC containment structures and foundations characterised by lower costs with respect to stainless steel structures;
- substitution of highly expensive corrugated steel liners with plane liners taking advantage of the geometric compensation of thermal dilations due to the conical shape of the tank;
- possibility of employing freezing passive systems for the concrete basement made of HPC, able to sustain temperature levels higher than those for OPC;
- fewer problems when the tank is located on low-strength soils.

The planned research activities required the upgrade of a F.E. coupled model for heat and mass transport (plus mechanical balance) to estimate concrete tanks durability under prolonged thermal loads and cyclic temperature variations due to changes in the salts level. The presence of a surrounding soil volume is additionally accounted for to evaluate environmental risk scenarios.

7.1 Mathematical-numerical modeling of concrete

Concrete is treated as a multiphase system where the voids of the skeleton are partly filled with liquid and partly with a gas phase (Baggio et al., 1995; Gawin et al., 1999). The liquid phase consists of bound water (or adsorbed water), which is present in the whole range of water contents of the medium, and capillary water (or free water), which appears when water content exceeds so-called solid saturation point S_{ssp} (Couture et al., 1996), i.e. the upper limit of the hygroscopic region of moisture content. The gas phase, i.e. moist air, is a mixture of dry air (non-condensable constituent) and water vapour (condensable gas), and is assumed to behave as an ideal gas.

The approach here is to start from a phenomenological model (Schrefler et al., 1989; Majorana et al., 1997; Majorana et al., 1998; Majorana & Salomoni, 2004 (b); Salomoni et al., 2007 (a)), originally developed by Bažant and co-authors, e.g. (Bažant, 1975; Bažant & Thonguthai, 1978; Bažant & Thonguthai, 1979; Bažant et al., 1988), in which mass diffusion and heat convection-conduction equations are written in terms of relative humidity, to an upgraded version in which its non-linear diffusive nature is maintained as well as the substitution of the linear momentum balance equations of the fluids with a constitutive equation for fluxes, but new calculations of thermodynamic properties for humid gases are implemented too to take into account different fluid phases as well as high ranges of both pressure and temperature. Additionally, Darcy's law is abandoned when describing gas flow through concrete.

The proposed model couples non-linear geometric relations with empirical relations; to enhance its predictive capabilities, a predictor-corrector procedure is supplemented to check the exactness of the solution. For additional details the reader is referred to (Salomoni et al., 2007 (b); Salomoni et al., 2008; Salomoni et al., 2009).

7.2 Numerical analyses

A conical tank for storing hot salts has been modelled through the F.E. research code NEWCON3D (Figure 16) using 330 8-node isoparametric elements (axis-symmetric condition). In agreement with the design criteria, it is proposed to employ a High Performance Concrete (HPC), particularly a C90 for this analysis, to increase both the operational temperature up to 120°C -against the usual 90°C for ordinary concretes- and concrete durability. The whole tank is composed by a flat stainless steel liner in contact with the salts and a ceramic fibre blanket (not modelled here) close to the concrete main structure (Figure 15). An additional passive cooling system is supposed to be added within the concrete thickness to reach such operational temperature on concrete surfaces. Geometric details have not been included for privacy reasons.

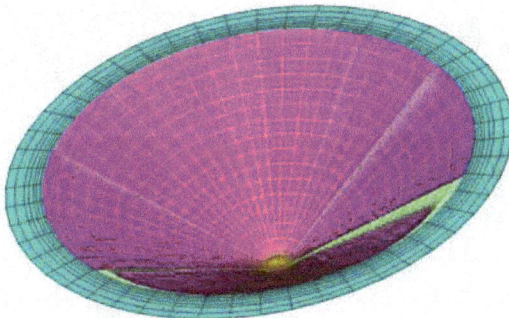

Fig. 16. F.E. discretization for the thermal storage concrete tank.

The adopted material properties are listed in Table 11.

Water/cement ratio	- 0.29
Elastic modulus [MPa]	$0.367 \cdot 10^5$
Poisson's ratio	0.18
Reference diffusivity along x/y directions [mm²/day]	$0.1 \cdot 10^2$
Intrinsic liquid permeability [mm²]	$2.0 \cdot 10^{-19}$
Unrestrained shrinkage for h = 0 (ε_{sh})	$-0.4 \cdot 10^{-2}$
Thermal expansion coefficient of solid	$0.12 \cdot 10^{-4}$
Hygro-thermal coefficient	$0.5 \cdot 10^{-2}$
Thermal capacity [N/(mm² K)]	2.0
Heat conductivity along x/y directions [N/(day K)]	$0.18144 \cdot 10^6$
Coefficient α_0 for diffusivity	$0.5 \cdot 10^{-1}$

Table 11. Material parameters for concrete C90.

The concrete tank is subjected to transient heating from the internal side assuming to reach the maximum temperature of 100°C in 8 days; the concrete tank has initially a relative humidity of 60% and a temperature of 30°C. In the first analyses (pushed up to about 4 months) the tank is supposed to be simply supported on its basement only.

The results in terms of R.H. (a) and temperature (b) are presented in Figure 17 (3D plot): the development of the R.H. bowl in time (along a typical tank section) is clearly evident; the peaks in R.H. for the zone closest to the heated surface are not referable to the phenomenon of "moisture-clog" (Majorana et al., 1998; Chung et al., 2006), because of the limited value of the temperature gradient, but anyway it is driven by the coupling between humidity and temperature fields and it is connected to the low intrinsic liquid permeability of the adopted HPC. Once 100°C has been reached, concrete starts depleting itself of water, thereby making the relative humidity values tend towards zero (but very slowly: in fact, after about 4 months, a concrete thickness of about 255 mm is still in saturated conditions).

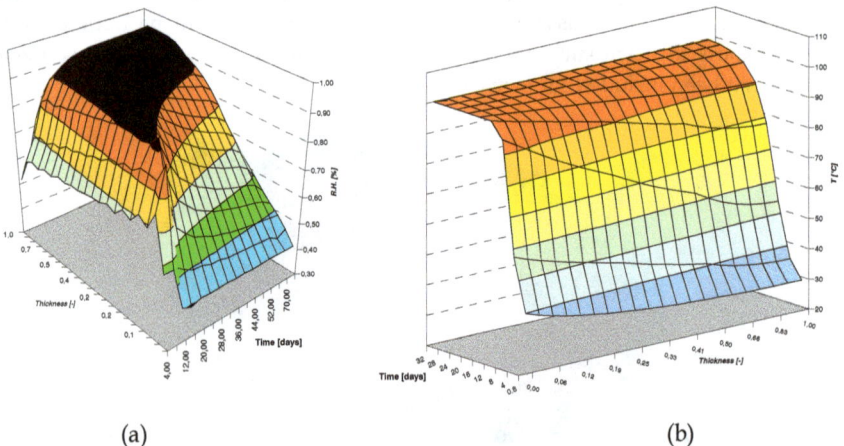

(a)   (b)

Fig. 17. R.H. (a) and temperature (b) time-history along a typical tank section.

The desaturation occurring within the concrete tank is mainly caused by the evaporation of water, resulting in formation of a zone of increased vapour pressure. Vapour pressure gradients cause vapour flow towards both the heated surface and the external side of the tank. Moreover, the existing temperature gradient causes thermo-diffusion of water vapour towards the colder layer of the wall. These vapour flows result in an increase in R.H. above its initial value as well as in condensation of vapour in the colder layers and subsequent slight increase in saturation.

The effect of a surrounding ground volume (dry sand) has been additionally evaluated on the development of the thermal front up to 2 years; the main results are here recalled (Salomoni et al., 2008): if considering a sand with a specific heat of 781.25 J/(kg K) and a thermal conductivity of 0.35 W/(m K), a shift the maximum temperature to farther times is revealed with respect to the results obtained from the analysis for the tank only. Additionally, the peak of 100°C seems to be not reachable even after several months.

The situation is clearly preferable when considering the durability performances of the concrete tank, but anyway the heat level is again not admissible for the soil: under such temperatures chemical reactions can take place if organic materials are present, even if a soil treatment is usually performed; moreover, a desiccation of zones around the tank combined with possible re-wettings due to rainfalls could induce relative displacements and consequently tank movements. Hence, being the hypothesis of an additional cooling system (or an additional foundation) too expensive in the design of such structures, it should be planned to use gravelly soils when preparing the surrounding embankment.

8. Conclusions

This Chapter gives a general view related to the last experiences R&D in the field of new technologies for solar energy exploitation within the Italian context, directly exportable abroad due to the followed design and analysis methodologies. The main structures and elements characterizing a solar power plant with parabolic-trough solar concentrators and a double-tank below ground system are studied, evidencing the fundamental design and modelling results.

From the process of analysis and verification of a 100 m parabolic trough solar concentrator, with reference to each system, subsystem and construction details, some remarks can be evidenced. Within this design approach, four operational states corresponding to different design winds, i.e. $W_1 = 7$ km/h, $W_2 = 14$ km/h, $W_3 = 21$ km/h e $W_4 = 28$ km/h, have been defined, to which two SLSs (for medium wind) and two ULSs (for peak wind) are respectively associated. Then, considering that the snow load acts simultaneously with the self-weight and with some probability contemporaneously with the wind load, the following conclusions can be stated:

- *Concentration system*: the adoption of a 3D FE model, corresponding to a 12 m module, has been considered as adequate to perform such a design and verification procedure. Safety or functioning problems have not been evidenced in the SLSs, linked to the operational states with W_1 and W_2 wind types. With reference to the elastic ULS (ELS), the maximum value for the σ_{eq} has been locally overcome (higher and middle part of the centering and some zones connecting the centering with the frames); such trend has been confirmed in the collapse ULSs.
- *Torque system* (*omissis*): the adoption of a simplified 3D FE model, in which the structure is characterized by a series of beams with sections of appropriate bending and torsion

inertias, has been considered as adequate to perform such a design and verification procedure. Safety, functioning and instability problems have not been evidenced in the SLSs, linked to the operational states with W_1 and W_2 wind types. In fact, with reference to the SLS verifications for weak wind, the only imposed constraint is that the maximum rotation of the torque tube is lower than 2 mrad and this constraint is largely satisfied, as well as the displacements requirements (limit of 20 mm for the camber). In relation to the ELS, the only constraint is the maintenance of the structural behaviour within the elastic limits in any condition, as proved by the obtained high safety factors. When the structure is verified under the collapse ULS, collapse must not occur in the standard safety position: two load cases exist for which such requirement is not satisfied and the critical action is given by the snow. Differently, the stability analyses referring to the heaviest load conditions (W_3 and W_4) show safety factors always higher than one. As regards the constructive details, the requirement of minimum distances between the connectors and the plates' edges, as indicated by the Eurocode 3, is not satisfied so that appropriate design prescriptions have been planned by the authors.

-   *Module's support system* (*omissis*): once again a simplified 3D FE model, in which the structure is characterized by a series of beams with sections of appropriate bending and torsion inertias, has been adopted. Both the intermediate and final supports, under SLSs for weak wind, are characterized by low stresses, as well as under SLSs for medium wind. The constraint of a structural behaviour within the elastic range in the elastic ULS is fully respected, as well as the absence of collapse in the standard safety position (collapse ULS). The stability analyses referring to the heaviest load conditions (W_3 and W_4) show again safety factors always higher than one.

A series of linear and non-linear static 3D analyses have been followed by modal, spectral and generally dynamic analyses to understand the structural behaviour of the whole CSP system and the first design sketch has been consequently updated. The analyses revealed an appropriate response of the whole structure under a variety of actions and once, e.g., the material strength had been locally overcome, local non-linear analyses have been conducted and the constructive details eventually re-designed.

Additionally, within the medium temperature field, an innovative approach has been presented for the conceptual design of liquid salts concrete storage systems. A multi-tank sensible-heat storage system has been proposed for storing thermal energy, with a two-tanks molten salt system. The hygro-thermal behaviour of a HPC tank has been assessed through a coupled F.E. code based on the theory by Bažant and enhanced by additional constitutive and thermodynamic relationships. A predictor-corrector procedure has been included to check the exactness of the solution. The study allows for estimating the durability performances of the tank: after about one month, all the structure is fully heated, possibly inducing thermal damage within concrete; such a result is slightly modified when modelling the domain more in detail, i.e. tank plus surrounding ground, or when changes in the salts level are considered. Even if at present some geometric and mechanical characteristics are still to be fixed, so that they can consequently induce an unavoidable uncertainty on the numerical results, the generality of the approach is not affected by such restrictions, and the results themselves can be evaluated as first guidelines in defining design criteria for liquid salts concrete systems. In fact, this study is the first step in a new research field and is being extended within the Italian Research Project "Elioslab – Research Laboratory for Solar Technologies at High Temperatures" started at the end of 2007.

Again, independently on the specificity and interest of the application, it has been shown here that the fully coupled mathematical-numerical model proposed (whose details have been reported for the hygro-thermal field only and whose predictive abilities have already been demonstrated in (Salomoni et al., 2007) even if in its not-upgraded form) has been enhanced through additional thermodynamic and constitutive relationships, allowing for obtaining more complete results in terms of water vapour pressure, gas pressure and capillary pressure which become fundamental variables mainly when higher temperature regimes are to be considered.

9. References

Antonaia, A.; Avitabile, M.; Calchetti, G.; Crescenzi, T.; Cara, G.; Giannuzzi, G.M.; Maccari, A.; Miliozzi, A.; Rufoloni, M.; Prischich, D.; Vignolini, M. (2001). Design sketch of the parabolic trough collector for solar plants. *ENEA/TM/PRES/2001_09*, Rome, Italy (technical report, in Italian).

Baggio, P.; Majorana, C.E.; Schrefler, B.A. (1995). Thermo-hygro-mechanical analysis of concrete. *International Journal for Numerical Methods in Fluids*, Vol. 20, 573-595.

Bažant, Z.P. (1975). Pore pressure, uplift, and failure analysis of concrete dams. *Int. Commission on Large Dams*, Swansee, UK.

Bažant, Z.P.; Thonguthai, W. (1978). Pore pressure and drying of concrete at high temperature. *Journal of the Engineering Materials Division*, ASME, Vol. 104, 1058-1080.

Bažant, Z.P.; Thonguthai, W. (1979). Pore pressure in heated concrete walls: theoretical predictions. *Magazine of Concrete Research*, Vol. 31, No.107, 67-76.

Bažant, Z.P.; Chern, J.C.; Rosenberg, A.M.; Gaidis, J.M. (1988). Mathematical Model for Freeze-Thaw Durability of Concrete. *Journal of the American Ceramic Society*, Vol. 71, No. 9, 776-83.

Chung, J.H.; Consolazio, G.R.; McVay, MC. (2006). Finite element stress analysis of a reinforced high-strength concrete column in severe fires. *Computers and Structures*, Vol. 84, 1338-1352.

Copeland, R.J.; West, R.E.; Kreith, F. (1984). Thermal Energy Storage at 900°C. *Proc. 19th Ann. Intersoc. Energy Conversion Engrg. Conf.*, San Francisco, Aug. 19-24, 1171-1175.

Couture, F.; Jomaa, W.; Ruiggali, J.R. (1996). Relative permeability relations: a key factor for a drying model. *Transport in Porous Media*, Vol. 23, 303-335.

ENEA (2003). http://www.enea.it/com/ingl/solarframe.htm.

Gawin, D.; Majorana, C.E.: Schrefler, B.A. (1999). Numerical analysis of hygro-thermal behaviour and damage of concrete at high temperature. *Mechanics of Cohesive-Frictional Materials*, Vol. 4, 37-74.

Giannuzzi, G.M.; Majorana, C.E.; Miliozzi, A.; Salomoni, V.A.L.; Nicolini, D. (2007) Structural design criteria for steel components of parabolic-trough solar concentrators. *Journal of Solar Energy Engineering*, Vol. 129, 382-390.

Herrmann, U.; Kelly, B.; Price, H. (2004). Two-tank molten salt storage for parabolic trough solar power plants. *Energy*, Vol. 29, No. 5-6, 883-893.

Ives, J.; Newcomb, J.C.; Pard, A.G. (1985). High Temperature Molten Salt Storage. *SERI/STR-231-2836* (technical paper).

Majorana, C.E.; Salomoni, V.; Secchi, S. (1997). Effects of mass growing on mechanical and hygrothermic response of three-dimensional bodies. *Journal of Materials Processing Technology*, PROO64/1-3, 277-286.

Majorana, C.E.; Salomoni, V; Schrefler, B.A. (1998). Hygrothermal and mechanical model of concrete at high temperature. *Materials and Structures*, Vol. 31, 378-386.

Majorana, C.; Salomoni, V. (2004) (a). Selection, elaboration and application of recommendations for designing parabolic trough solar concentrators. Functional description, classification and selection of design codes for their structural elements. *Report ENEA-TRASTEC*, Rome-Padua, Italy (in Italian).

Majorana, C.; Salomoni, V. (2004) (b). Parametric analyses of diffusion of activated sources in disposal forms. *Journal of Hazardous Materials*, A113,. 45-56.

Majorana, C.; Salomoni, V. (2005) (a). Design guide for parabolic trough solar concentrators. *Report ENEA-TRASTEC*, Rome-Padua, Italy (in Italian).

Majorana, C.; Salomoni, V. (2005) (b). Analyses and structural verifications for a 100 m-parabolic trough solar concentrator. *Report ENEA-TRASTEC*, Rome-Padua, Italy (in Italian).

Miliozzi, A.; Nicolini, D.; Giannuzzi, G.M. (2007). Evaluation of wind action on parabolic trough concentrators for a high temperature solar plant. *Enea Report RT/2007/13/TER* (in Italian).

Pilkington Solar International GmbH (2000). Survey of Thermal Storage for Parabolic-Trough Power Plants, *NREL/SR-550-27925* (technical report).

Rubbia, C.; and ENEA Working Group (2001). Solar thermal energy production: guidelines and future programmes of ENEA. *ENEA/TM/PRES/2001_7*, Rome, Italy (technical report).

Salomoni, V.A.; Giannuzzi, G.M.; Majorana, C.E.; Miliozzi, A.; Nicolini, D. (2006). Structural design of parabolic-trough solar concentrators' steel components against wind and natural hazards, *Proceedings of the 3^ Int. Conf. on Protection of Structures Against Hazards* (Majorana, Salomoni, Lok Eds.), Venice, Italy, sept. 28-29 (ISBN 981-05-5561-X) (keynote lecture).

Salomoni, V.A.; Mazzucco, G.; Majorana, C.E. (2007) (a). Mechanical and durability behaviour of growing concrete structures. *Engineering Computations*, Vol. 24, No. 5, 536-561.

Salomoni, V.A.; Majorana, C.E.; Khoury, G.A. (2007) (b). Stress-strain experimental-based modeling of concrete under high temperature conditions. In: B.H.V. Topping (Ed.), *Civil Engineering Computations: Tools and Techniques*, Ch. 14, Saxe-Coburg Publications, 319-346.

Salomoni, V.A.; Majorana, C.E.; Giannuzzi, G.M.; Miliozzi, A. (2008) Thermal-fluid flow within innovative heat storage concrete systems for solar power plants. *International Journal of Numerical Methods for Heat and Fluid Flow* (Special Issue), Vol. 18(7/8), 969-999.

Salomoni, V.A.; Majorana, C.E.; Mazzucco, G.; Xotta, G.; Khoury, G.A. (2009). Multiscale Modelling of Concrete as a Fully Coupled Porous Medium. In: J.T. Sentowski (Ed.), *Concrete Materials: Properties, Performance and Applications*, Ch. 3, NOVA Publishers, 2009 (in press).

Schrefler, B.A.; Simoni, L.; Majorana, C.E. (1989). A general model for the mechanics of saturated-unsaturated porous materials. *Materials and Structures*. Vol. 22, 323-334.

Amorphous Silicon Carbide Photoelectrode for Hydrogen Production from Water using Sunlight

Feng Zhu[1], Jian Hu[1], Ilvydas Matulionis[1], Todd Deutsch[2], Nicolas Gaillard[3], Eric Miller[3], and Arun Madan[1]

[1]*MVSystems, Inc., 500 Corporate Circle, Suite L, Golden, CO, 80401*
[2]*Hawaii Natural Energy Institute (HNEI), University of Hawaii at Manoa, Honolulu, HI 96822,*
[3]*National Renewable Energy Laboratory (NREL), Golden, CO 80401, USA*

1. Introduction

Hydrogen is emerging as an alternative energy carrier to fossil fuels. There are many advantages of hydrogen as a universal energy medium. For example, it is non-toxic and its combustion with oxygen results in the formation of water to release energy. In this chapter, we discuss the solar to hydrogen production directly from water using a photoelectrochemical (PEC) cell; in particular we use amorphous silicon carbide (a-SiC:H) as a photoelectrode integrated with a-Si tandem photovoltaic (PV) cell. High quality a-SiC:H thin film with bandgap ≥ 2.0eV was fabricated by plasma enhanced chemical vapor deposition (PECVD) technique using SiH_4, H_2 and CH_4 gas mixture. Incorporation of carbon in the a-SiH film not only increased the bandgap, but also led to improved corrosion resistance to an aqueous electrolyte. Adding H_2 during the fabrication of a-SiC:H material could lead to a decrease of the density of states (DOS) in the film. Immersing the a-SiC:H(p)/a-SiC:H(i) structure in an aqueous electrolyte showed excellent durability up to 100 hours (so far tested); in addition, the photocurrent increased and its onset shifted anodically after 100-hour durability test. It was also found that a SiO_x layer formed on the surface of a-SiC:H, when exposed to air led to a decrease in the photocurrent and its onset shifted cathodically; by removing the SiO_x layer, the photocurrent increased and its onset was driven anodically. Integrating with a-Si:H tandem cell, the flat-band potential of the PV/a-SiC:H structure shifts significantly below the H_2O/O_2 half-reaction potential and is in an appropriate position to facilitate water splitting and has exhibited encouraging results. The PV/a-SiC:H structure produced hydrogen bubbles from water splitting and exhibited good durability in an aqueous electrolyte for up to 150 hours (so far tested). In a two-electrode setup (with ruthenium oxide as counter electrode), which is analogous to a real PEC cell configuration, the PV/a-SiC:H produces photocurrent of about 1.3 mA/cm^2 at zero bias, which implies a solar-to-hydrogen (STH) conversion efficiency of over 1.6%. Finally, we present simulation results which indicate that a-SiC:H as a photoelectrode in the PV/a-SiC:H structure could lead to STH conversion efficiency of >10%.

2. Principles and status of using semiconductor in PEC

In general, hydrogen can be obtained electrolytically, photo-electrochemically, thermochemically, and biochemically by direct decomposition from the most abundant material on earth: water. Though a hydrogen-oxygen fuel cell operates without generating harmful emissions, most hydrogen production techniques such as direct electrolysis, steam-methane reformation and thermo-chemical decomposition of water can give rise to significant greenhouse gases and other harmful by-products. We will briefly review the solid-state semiconductor electrodes for PEC water splitting using sunlight. Photochemical hydrogen production is similar to a thermo-chemical system, in that it also employs a system of chemical reactants, which leads to water splitting. However, the driving force is not thermal energy but sunlight. In this sense, this system is similar to the photosynthetic system present in green plants. In its simplest form, a photoelectrochemical (PEC) hydrogen cell consists of a semiconductor as a reaction electrode (RE) and a metal counter electrode (CE) immersed in an aqueous electrolyte, and PEC water splitting at the semiconductor-electrolyte interface drove by sunlight, which is of considerable interest as it offers an environmentally "green" and renewable approach to hydrogen production (Memming, 2000).

2.1 Principles of PEC

The basic principles of semiconductor electrochemistry have been described in several papers and books (Fujishima & Honda, 1972; Gerscher & Mindt, 1968; Narayanan & Viswanathan, 1998; Memming, 2000; Gratzel, 2001).

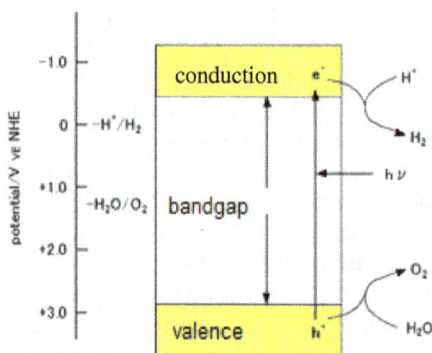

Fig. 1. The band diagram of the PEC system. The conduction band edge needs to be located negative (on an electrochemical scale) high above the reduction potential of water, the valence band edge positive enough below the oxidation potential of water to enable the charge transfer. NHE stands for "normal hydrogen electrode".

The only difference between a photoelectrochemical and a photovoltaic device is that in the PEC case, a semiconductor-electrolyte junction is used as the active layer instead of the solid-state junctions in a photovoltaic structure. In both cases, a space charge region is formed where contact formation compensates the electrochemical potential differences of electrons on both sides of the contact. The position of the band edges of the semiconductor at the interface can be assumed in a first approximation to be dependent only on the pH of the solution and independent of the potential (Fermi level) of the electrode or the electrolyte (Memming, 2000; Kuznetsov & Ulstrup, 2000). For direct photoelectrochemical decomposition of water, several primary requirements of the semiconductor must be met: the semiconductor system must generate sufficient voltage (separation of the quasi Fermi levels under illumination) to drive the electrolysis, the energetic of the semiconductor must overlap that of the hydrogen and oxygen redox reactions (saying the band positions at the semiconductor-electrolyte interface have to be located at an energetically suitable position as shown in Fig.1), the semiconductor system must be stable in aqueous electrolytes, and finally the charge transfer from the surface of the semiconductor must be fast enough not only to prevent corrosion but also reduce energy losses due to overvoltage (Gerscher & Mindt, 1968; Narayanan & Viswanathan, 1998; Memming, 2000).

Neglecting losses, the energy required to split water is 237.18 kJ/mol, which converts into 1.23 eV, i.e. the PV device must be able to generate more than 1.23 Volts. The STH conversion efficiency in PEC cells can be generally expressed as

$$\text{Efficiency} = \frac{\text{chemical energy in hydrogen produced in a PEC cell}}{\text{energy in the sunlight over the collection area}} = \frac{J_{ph}V_{Ws}}{E_S} \quad (1)$$

where J_{ph} is the photocurrent density (in mA/cm^2) generated in a PEC cell, $V_{WS} = 1.23$ V is the potential corresponding to the Gibbs free energy change per photon required to split water, and Es is the solar irradiance (in mW/cm^2). Under AM1.5 G illumination, a simple approximation for the STH efficiency is J_{ph} times 1.23(in %) (Memming, 2000; Miller & Rocheleau, 2002).

2.2 Status of using semiconductor in PEC

Although as early as in 1839 E. Becquerel (Memming, 2000) had discovered the photovoltaic effect by illuminating a platinum electrode covered with a silver halide in an electrochemical cell, the foundation of modern photoelectrochemistry has been laid down much later by the work of Brattain and Garret and subsequently Gerischer (Bak, et al., 2002; Mary & Arthru, 2008), who undertook the first detailed electrochemical and photoelectrochemical studies of the semiconductor–electrolyte interface. From then on, various methods of water splitting have been explored to improve the hydrogen production efficiency. So far, many materials that could be used in the PEC cell structure have been identified as shown in Fig.2. However, only a few of the common semiconductors can fulfil the requirements presented above even if it is assumed that the necessary overvoltage is zero. It should be noted that most materials have poor corrosion resistance in an aqueous electrolyte and posses high bandgap, which prevents them from producing enough photocurrent (Fig.5).

Photoelectrolysis of water, first reported in the early 1970's (Fujishima, 1972), has recently received renewed interest since it offers a renewable, non-polluting approach to hydrogen production. So far water splitting using sunlight has two main approaches. The first is a two-step process, which means sunlight first transform into electricity which is then used to split water for hydrogen production (Tamaura, et al., 1995; Hassan & Hartmut, 1998). Though only about 2V is needed to split water, hydrogen production efficiency depends on large current via wires, resulting in loss due to its resistance; the two-step process for hydrogen production is complex and leads to a high cost.

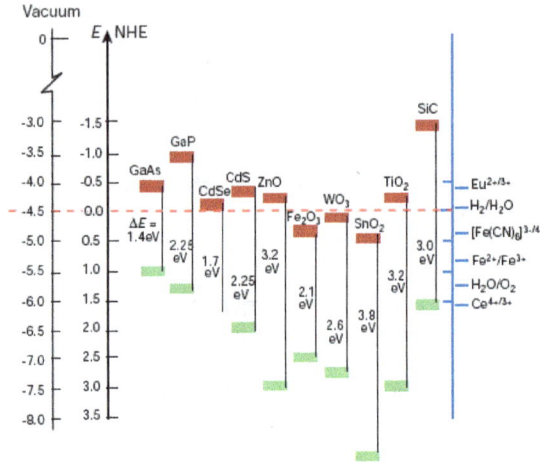

Fig. 2. Band positions of some semiconductors in contact with aqueous electrolyte at pH1. The lower edge of the conduction band (red colour) and the upper edge of the valence band (green colour) are presented along with the bandgap in electron volts. For comparison, the vacuum energy scale as used in solid state physics and the electrochemical energy scales, with respect to a normal hydrogen electrode (NHE) as reference points, are shown as well as the standard potentials of several redox couples are presented against the standard hydrogen electrode potential on the right side (Gratzel, 2001).

Another approach is a one-step process, in which there are no conductive wires and all the parts are integrated for water splitting, as shown in Fig.3. In this structure as there are no wires, hence no loss. Another advantage is that the maintenance is low compared to the two-step process discussed above.

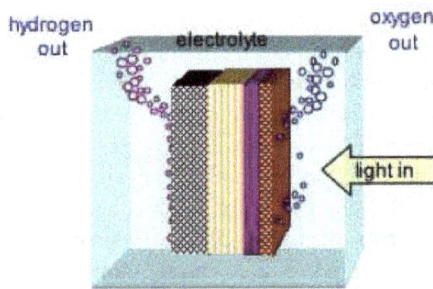

Fig. 3. Generic Planar Photoelectrode Structure with Hydrogen and Oxygen Evolved at Opposite Surfaces (Miller & Rocheleau, 2002)

In 1972, Fujishima and Honda used n type TiO_2 as the anode and Pt as the cathode to form the PEC structure and achieved 0.1% of STH efficiency (Fujishima & Honda, 1972). In this system TiO_2 absorbed the sunlight to produce the current while its bandgap (~3.2eV) provided the needed voltage for water splitting. Although TiO_2 is corrosion resistant in an aqueous electrolyte, but because of its high band gap leads to absorption of sunlight in the short wavelength range only, resulting in a small current and hence a low STH efficiency. In order to increase the current, some researchers are attempting to narrow its bandgap to enhance its absorption, and with limited success (Masayoshi, et al., 2005; Nelson & Thomas, 2005; Srivastava, et al. 2000).

In 1975, Nozik first reported using $SrTiO_3(n)$ and GaP(p) photoelectrodes as the anode and cathode respectively (Nozik, 1975) and obtained a STH efficiency of 0.67%. In 1976, Morisaki's group introduced utilizing a solar cell to assist the PEC process for hydrogen production (Morisaki, et al., 1976). Silicon solar cell was integrated with TiO_2 in series to form a PEC system, which exhibited higher photo current by absorbing more sunlight and higher voltage. Later, Khaselev and Turner in 1998 reported 12.4% of STH efficiency using p-GaInP$_2$/n-GaAs/p-GaAs/Pt structure (Khaselev & Turner, 1998); in this, surface oxygen was produced at the p-GaInP$_2$ side and hydrogen from the Pt side. Although this structure exhibited high STH efficiency, the corrosion resistance of p-GaInP$_2$ in an aqueous electrolytes was very poor, and was almost all etched away within a couple of hours. (Deutsch et al., 2008).

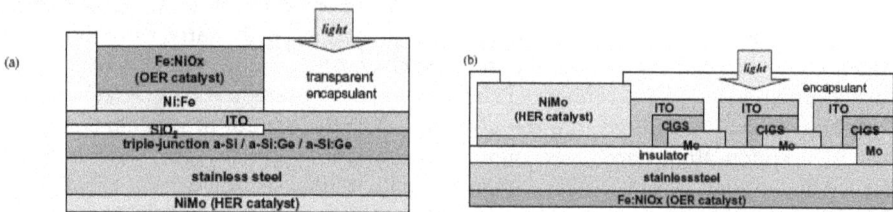

Fig. 4. (a) A-Si triple PV junctions and (b) CIGS PV cell integrated into a PEC system (Miller, et al., 2003)

Richard et al., reported 7.8% of STH efficiency by using NiMo or CoMo as cathode, Ni-Fe-O metal as anode and integrating with a-Si/a-Si:Ge/a-Si:Ge triple junctions solar cell as shown in Fig.4 (a) (Richard, et al., 1998). They also used copper indium gallium selenide (CIGS) module to replace a-Si triple junctions to produce even higher photo current as shown in Fig.4 (b).

Yamada, et al., also used a similar structure (Co–Mo and the Fe–Ni–O as the electrodes) and achieved 2.5% STH efficiency (Yamada, et al., 2003). More notably, a STH efficiency of 8% was reported by Lichta, et al., using AlGaAs/Si RuO$_2$/Pt black structure (Lichta, et al., 2001). In this structure, solar cell was separated from the aqueous electrolyte to avoid being corroded; it should be noted that the fabrication process for the device was very complicated. The non-transparent electrode had to cover the active area of the solar cell in order to enlarge electrode-electrolyte contact to as large area as possible.

In 2006, a "hybrid" PEC device consisting of substrate/amorphous silicon (nipnip)/ ZnO/WO$_3$, which would lead to ~3% solar-to-hydrogen (STH) conversion efficiency, was reported (Stavrides, et al., 2006). In this configuration, transparent WO$_3$ prepared by sputtering technique acted as the photoelectrode, whereas the amorphous silicon tandem solar cell was used as a photovoltaic device to provide additional voltage for water splitting at the interface of photoelectrode-electrolyte. In this structure, primarily the UV photons are absorbed by WO$_3$ while the green to red portion of the AM1.5 Global spectra was absorbed in the a-Si tandem photovoltaic device. Due to a high bandgap (Eg) (2.6-2.8eV) of the WO$_3$ photoelectrode, the photocurrent density of this hybrid PEC device is limited to no more than 5 mA/cm^2 (as shown in Fig.5), resulting in low STH efficiency.

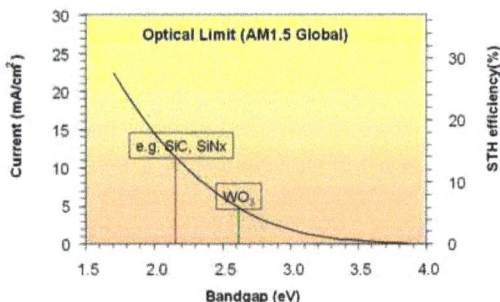

Fig. 5. Maximum current available as a function of the bandgap (E_g) of various materials under Global AM1.5 illumination (assumptions are that all the photons are absorbed for energy in excess of the band gap and the resulting current is all collected)

The US Department of Energy has set a goal to achieve STH conversion efficiency of 10% by 2018 (Miller & Rocheleau, 2001). To reach this goal, a photocurrent > 8.1mA/cm^2 is needed in PEC devices as deduced from equation (1). As shown in Fig.5, materials with narrower bandgap could produce higher photo current, such as a-SiC:H and a-SiN$_x$:H which can be routinely grown using plasma enhanced chemical vapor deposition (PECVD) technique and their bandgaps can be tailored into the ideal range by the control of stoichiometry, i.e., $\leq$ 2.3eV. In addition to generating enough photocurrent, necessary for STH conversion efficiency higher than 10%, a-SiC:H when in contact with the electrolyte, could also produce a significant photovoltage as other semiconductors (Nelson&Thomas, 2008), which would then reduce the voltage that is needed from the photovoltaic junction(s) for water splitting. Further, incorporation of carbon should lead to a more stable photoelectrode compared to pure amorphous silicon, which has poor resistance to corrosion when in contact with the electrolyte (Mathews, et al., 2004; Sebastian, et al., 2001).

3. a-SiC:H materials and its application as absorber layer in solar cells

A-SiC:H films were fabricated in a PECVD cluster tool system specifically designed for the thin film semiconductor market and manufactured by MVSystems, Inc. The intrinsic a-SiC:H films were deposited using CH$_4$, SiH$_4$, and H$_2$ gas mixtures at 200°C substrate temperature. The detail deposition parameters were presented in the reference [Zhu, et al., 2009].

3.1 a-SiC:H materials prepared by RF-PECVD
Fig.6 shows the bandgap (E_g), photoconductivity (σ_{ph}) and gamma factor (γ) as function of CH$_4$/(SiH$_4$+CH$_4$) gas ratio used during a-SiC:H growth. As CH$_4$/(SiH$_4$+CH$_4$) gas ratio increases, Eg increases from ~1.8eV to over 2.0eV (Fig.6 (a)) while the dark conductivity (σ_d) decreases to <1.0 x 10^{-10} S/cm (not shown here), which is the limit of the sensitivity of our measurement technique. We also note that σ_{ph} decreases from about 1.0 x 10^{-5} to 1.0 x 10^{-8} S/cm when CH$_4$/(SiH$_4$+CH$_4$) gas ratio increases.

Here, the γ, is defined from $\sigma_{ph} \propto F^\gamma$, where σ_{ph} is the photoconductivity and F is the illumination intensity; we infer the density of defect states (DOS) of the amorphous semiconductor from this measurement (Madan & Shaw, 1988). High-quality a-Si materials

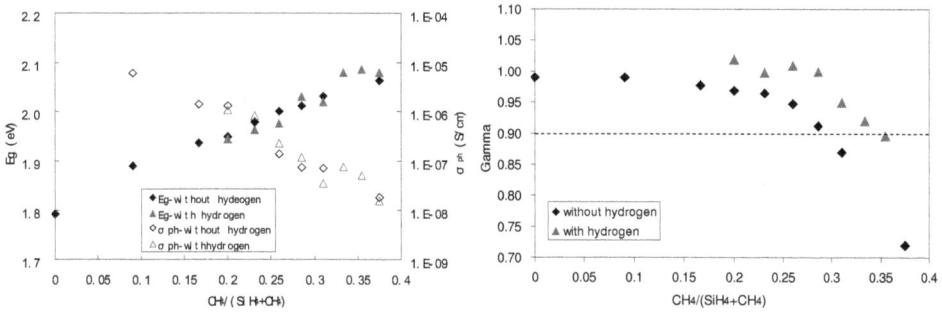

Fig. 6. (a) Bandgap (E_g), photoconducitivty (σ_{ph}) and (b) γ are plotted as function of the $CH_4/(SiH_4+CH_4)$ gas ratio used during the fabrication of a-SiC:H materials with/without H_2.

generally exhibits $\gamma > 0.9$. As shown in Fig.6 (b), when $CH_4/(SiH_4+CH_4)$ gas ratio is > 0.35, γ decreases to a low value of ~0.7, indicative of a material with high defect states. For $CH_4/(SiH_4+CH_4)$ gas ratio < 0.3, $\gamma > 0.9$, which indicates the DOS in materials is low. Fig.6 also shows the effect of hydrogen on a-SiC:H films. E_g and σ_{ph} of a-SiC:H films prepared with 100sccm H_2 flow during deposition process have similar value as that prepared without H_2 as shown in Fig.6 (a). It should be noted that use of H_2 during fabrication led to an increase of γ, as shown in Fig.6 (b), indicates that the DOS in the film is decreased due to removal of weak bonds due to etching and passivation (Yoon, et al., 2003; Hu, et al., 2004;).

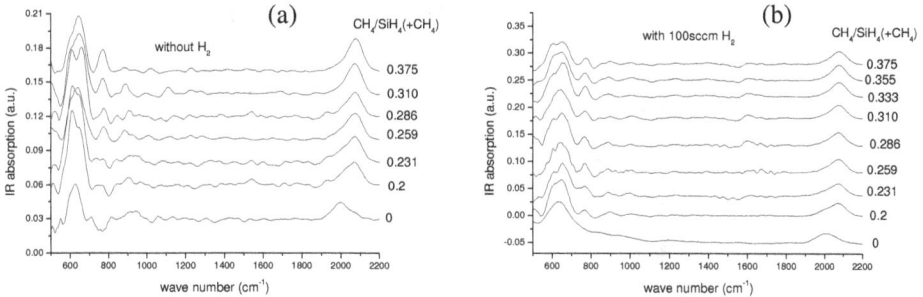

Fig. 7. IR spectra of a-SiC:H as function of $CH_4/(SiH_4+CH_4)$ gas ratio without (a) and with H_2 dilution (b).

Evidence of carbon incorporation in the films can be discerned from infrared (IR) spectroscopy. As shown in Fig.7, regardless of H_2 addition during deposition process, the peak at 2000 cm^{-1} (related to Si-H stretching vibration mode) always shifts towards 2080 cm^{-1}, once $CH_4/(SiH_4+CH_4)$ gas ratio is greater than 0.2. This shift of Si-H stretching vibration mode is mainly caused by incorporation of C atoms, and probably due to the back-bonding of the Si atoms to carbon (Hollingsworth, et al., 1987). In addition, it is found that in the a-SiC:H films using 100sccm H_2 dilution, the ratio of the absorption peak at 2080 cm^{-1} to 2000 cm^{-1} is smaller than that without the use of H_2 dilution, implying less defective films (Hu, et al., 2004). It is also seen from Fig.7 (a) that in the a-SiC:H without H_2 dilution, the IR peak at 780 cm^{-1}, which is related to Si-C stretching mode, increases as the CH$_4$/(SiH$_4$+CH$_4$) gas ratio increases; whereas films produced with 100sccm H$_2$ dilution, this peak keeps almost constant (as shown Fig.7(b)). This is likely due to the removal of SiH$_2$ and a decrease of carbon clusters in the films (Desalvo, et al., 1997). It was also found that at a fixed CH$_4$/(SiH$_4$+CH$_4$) =0.2, σ_{ph} is enhanced from 4.0 x 10^{-7} S/cm to 3.2 x 10^{-6} S/cm as the H$_2$ flow increased from 0 to 150sccm.

The decrease in σ_{ph} with increasing CH$_4$/(SiH$_4$+CH$_4$) gas ratio as shown in Fig. 6(a) is unlikely be due to an increase of the recombination centers related to defects since the γ factor is >0.9.The decrease of σ_{ph} results from a reduction in the absorption coefficient as Eg increases. In order to further evaluate this, the nominal photocurrent, Ip, at certain wavelength, under uniform bulk absorption (here we select wavelength 600nm) can be measured and the photocurrent be expressed as,

$$I_p = e.N_{ph\,(\lambda)}\,(1-R_\lambda)\,[1 - \exp\,(\,-\alpha_\lambda\,d)]\eta\bar{\imath}/t_t \qquad (2)$$

Where, $N_{ph(\lambda)}$ is the photon flux, R_λ is the reflection coefficient, α_λ the absorption coefficient, d the film thickness, η is the quantum efficiency of photo generation, $\bar{\imath}$ is the recombination lifetime and t_t is the transit time. Assuming that η, $\bar{\imath}$, t_t and $(1-R_\lambda)$ are constant for different films (i.e. different E_g), then to the first order approximation, the normalized photocurrent, $I_p/[1- \exp(-\alpha_\lambda d)]$, can account for the changes in the absorption coefficient as E_g varies (Madan & Shaw, 1988). It was indeed shown that the normalized photocurrent does not change significantly as E_g increases (Hu, et al., 2008). This is in contrast to the decrease in σ_{ph} with E_g as shown in Fig.6 (a), suggesting a low DOS, consistent with high γ (>0.9) throughout the range.

3.2 Photothermal deflection spectroscopy (PDS) spectrum of a-SiC:H films

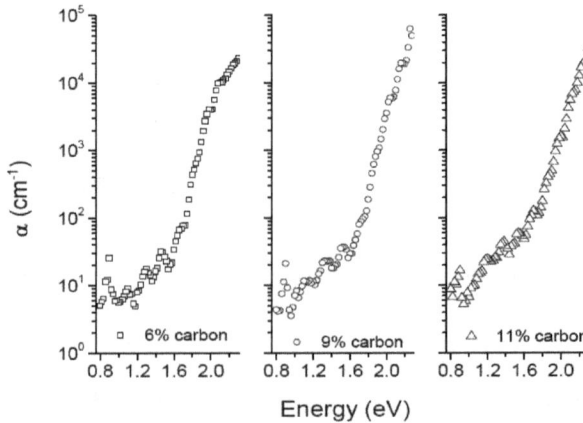

Fig. 8. Absorption coefficient curves of three a-SiC:H films, with differing carbon concentrations, measured using photothermal deflection spectroscopy (PDS)

Fig.8 shows the absorption coefficient of the three chosen films with differing carbon concentrations, prepared with H$_2$ dilution, measured by the photothermal deflection spectroscopy (PDS). Using energy dispersive x-ray spectroscopy on a JEOL JSM-7000F field emission scanning electron microscope with an EDAX Genesis energy dispersive x-ray spectrometer their carbon concentrations are 6, 9, and 11% (in atomic), corresponding to methane gas ratio, used in the fabrication, of 0.20, 0.29 and 0.33 respectively. The signal seen here is a convolution of optical absorption from every possible electronic region including extended, localized and deep defect states. In the linear region between about 1.7–2.1eV, the absorption coefficient primarily results from localized to extended state transitions and is known as the Urbach tail. This region can be described by $\alpha = \alpha_0 \exp(E/E_0)$ where E is the excitation energy and E_0 is the Urbach energy which is the inverse slope of the data when plotted versus $\ln(\alpha)$. Since the absorption coefficient here directly depends on the density of localized states, E_0 is considered to be a measure of the amount of disorder (Cody, et al., 1981). Their bandgap values are presented in Fig.6 (a) as previously discussed and E_0 is 78, 85, and 98 meV for carbon concentrations of 6, 9, and 11%, respectively. For comparison, a typical value for device grade a-Si:H is ~50 meV(Madan & Shaw, 1988). As the carbon concentration increases, so too does the value of E_0. This is expected as the density of localized states is increasing with more disorder created by introducing more carbon. Also, there is an increase in the bandgap from E_{04} = 2.06-2.18eV with carbon concentration (E_{04} is defined as the energy value where the absorption coefficient $\alpha = 10^4$ cm^{-1}). This is known to be a result of at least some of the carbon being incorporated in the form of sp^3 carbon which is essentially an insulator (Solomon, 2001). The feature at 0.88eV in Fig. 8 is an overtone of an O-H vibrational stretch mode from the quartz substrate.

As the bandgap increases with carbon incorporation, as evidenced from the PDS data, the Urbach energies are 50% to 100% higher than is typically seen in device grade a-Si:H. This is typically interpreted as an increase in localized states within the bandgap region just above the valence band and below the conduction band resulting from structural disorder. It is believed that the carbon is incorporated into our films as a mixture of sp^2 and sp^3 carbon from ESR test (Solomon, 2001; Simonds (a), et al., 2009).

3.3 a-SiC single junction devices

The previous results suggest that high quality a-SiC:H can be fabricated with Eg ≥2.0eV. To test the viability of a-SiC:H material in device application, we have incorporated it into a p-i-n solar cell in the configuration, glass/Asahi U-Type SnO$_2$:F/p-a-SiC:B:H/i-a-SiC:H/n-a-Si/Ag as shown in Fig.9. The Ag top contact defines the device area as 0.25cm^2. The thickness of i-layer is ~300nm.

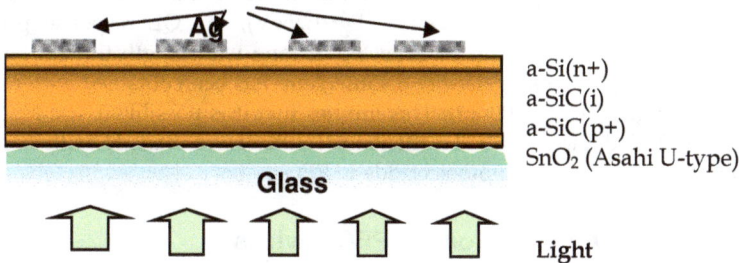

a-Si(n+)
a-SiC(i)
a-SiC(p+)
SnO$_2$ (Asahi U-type)

Glass

Light

Fig. 9. Configuration of p-i-n single junction solar cell

Three a-SiC:H i-layers with different carbon concentration were used in single junction solar cells. Fig.10 (a) and (b) show their J-V and quantum efficiency (QE) curves, respectively. As mentioned above, the three films with carbon concentration of 6%, 9%, and 11%, correspond to bandgaps of approx.2.0eV, 2.1eV and 2.2eV respectively. Though the bandgap increases with carbon concentration, the performance of a-SiC devices deteriorates quickly, especially the fill factor, implying an increase of the defects from carbon inclusion. As the carbon concentration increases, the QE peak shifts toward the short wavelength region   and becomes smaller (Fig. 10 (b)), resulting from higher defects density with bandgap (Madan & Shaw, 1988). The influence of defects resulting from increased carbon can also be seen in the dark J-V curves. Here carrier transport is only affected by the built-in field and the defects in the films. As the carbon concentration increases, the diode quality factor deduced from the dark J-V curves also increases, which also implies loss due to increased defect densities (Simonds (b), et al., 2009). The device performances variation is consistent with the PDS data discussed above, where E_0 is 78, 85, and 98 meV for carbon concentrations of 6, 9, and 11%, respectively; increasing E_0 is indicative of increased defect state density (Madan & Shaw, 1988).

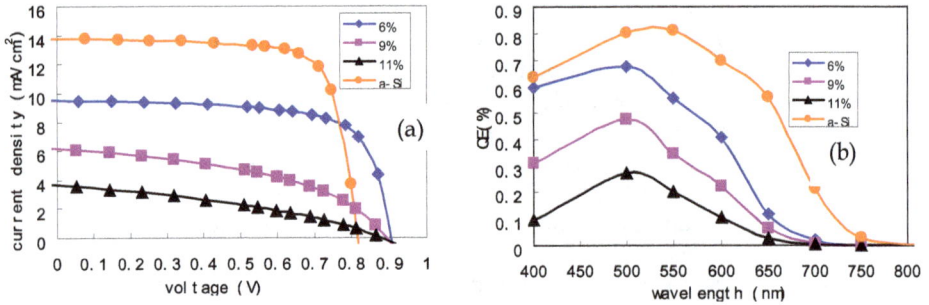

Fig. 10. (a) Illuminated J-V characteristics and (b) quantum efficiency (QE) curve of a-SiC:H single junction solar cell with different C concentrations(% in atomic) as labelled in the inset: for comparison purposes we have also included a-Si H device without any carbon in absorber layer.

Device using a-SiC:H with bandgap of 2.0eV exhibited a good performance. under AM1.5 illumination, with Voc = 0.91V, Jsc=11.64mA/cm^2, fill factor (FF) =0.657. We have also observed that FF under blue (400nm) and red (600nm) illumination exhibited 0.7 (not shown here), which indicates it is a good device and that a-SiC:H material is of high-quality. Compared with the normal a-Si:H devices (Eg~1.75eV), the QE response peak shifts towards a shorter wavelength; asis to be expected at long wavelength the QE response is reduced due to the increase in its Eg. Jsc of ~8.45mA/cm^2 has been obtained with reduced a-SiC:H intrinsic layer thickness (~100nm). This implies that it is possible to use a-SiC:H as a photoelectrode in PEC devices for STH efficiency >10%. Here a-SiC:H with bandgap of 2.0eV is selected to be used as the photoelectrode in PEC.

4. a-SiC:H used as a photoelectrode in PEC devices

An intrinsic a-SiC:H (~200nm) and a thin p-type a-SiC:H:B layer (~20nm) was used as the photoelectrode (Fig.11) to form a PV(a-Si tandem cell)/a-SiC:H device. In general, the a-

SiC:H behaves as a photocathode where the photo generated electrons inject into the electrolyte at the a-SiC:H/electrolyte interface to reduce the H^+ for hydrogen evolution. This way, anodic reaction and thus corrosion on a-SiC:H layer can be mitigated.

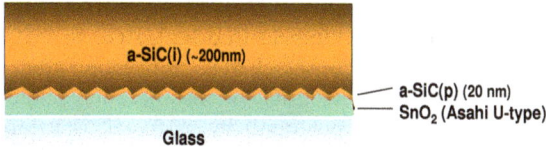

Fig. 11. Configuration of a-SiC:H photoelectrodes.

The current-potential characteristics of a-SiC:H photoelectrodes were measured with a three-electrode setup, with either saturated calomel electrode (SCE) or Ag/AgCl as the reference electrode (RE) and Pt as the counter electrode (CE). The samples were illuminated through the intrinsic a-SiC:H side under chopped light, using either a xenon or tungsten lamp (both calibrated to Global AM1.5 intensity calibrated with reference cell). However, due to difference in spectrum, the photocurrent of photoelectrodes varied depending on the light source used (Murphy, et al., 2006). Typical current-potential characteristics of a-SiC:H photoelectrodes are shown in Fig.12. The photocurrent density (J_{ph}) is defined as the difference between the current density without illumination (J_{dark}) and the current density under illumination. Initial experiments, using aqueous 0.0–0.5pH sulfuric acid electrolyte led to significant degradation during 10-minute test. The analysis of the initial results pointed towards using less acidic solutions (higher pH), which was described in elsewhere (Matulionis, et al., 2008). It was found that a better photoelectrode performance (diminished corrosion and higher J_{ph}) was achieved in a pH2 electrolyte.

Durability tests were carried out at NREL which involved a constant current density of -3 mA/cm² applied to the a-SiC:H photoelectrode. A tungsten lamp was used as the light source (calibrated to Global AM1.5 intensity with a 1.8eV reference cell). Electrolyte used was pH2 sulphamic acid/potassium biphthalate solution and a Triton X-100 surfactant.

4.1 Durability of a-SiC:H photoelectrodes

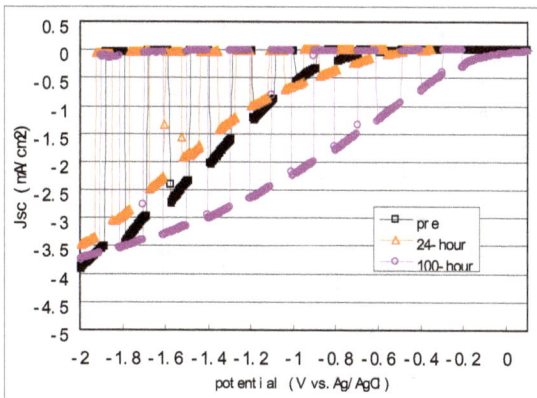

Fig. 12. J-V curves before and after 24- and 100-hour durability test on textured SnO₂ substrate

Fig.12 shows J-V curves of a-SiC:H photoelectrode on textured SnO_2 substrate before (blue curve) and after 24- (orange curve) and 100-hour (pink curve) durability tests. It is seen that after the 100-hour durability test, J_{dark} remains very low, suggesting the a-SiC:H photoelectrode is stable in electrolyte up to 100-hour (so far tested). Photo images of the surface morphology of tested a-SiC:H photoelectrodes also verify that they remain largely intact. Interestingly, both J_{ph} and its onset change noticeably after the durability test, particularly after the 100-hour durability test. For instance, the photocurrent onset shifts anodically (towards a lower absolute potential) by ~0.6V. This means the extra voltage needed to overcome various overpotential and the non-ideal energy band edge alignment to the H_2O/O_2 redox potential is lower after the durability test. We noted that increase in the test-duration time resulted in larger onset shift and increase in J_{ph} . The exact reason for such a behavior is not known at the moment. Possible causes could be, (i) the native-grown SiO_x on the surface of a-SiC:H film is probably eliminated (as described in more details later), and (ii) modification of the surface of a-SiC:H photoelectrode.

Fig.13 shows the surface morphology of the a-SiC:H photolelectrode on Asahi U type SnO_2 before and after the 100-hour durability test. One can see that after the test, the surface

Fig. 13. SEM images of surface morphology of a-SiC:H photoelectrode (a) before and (b)after 100-hour durability test .

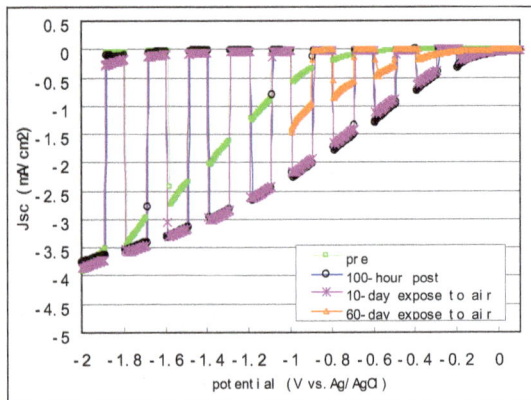

Fig. 14. J-V curves of a-SiC:H photoelectrode before and after 100-hour durability test, and then exposed to air for 10- and 60-day and tested.

morphology of a-SiC:H looks similar. The only difference is that after the test there are many tiny motes on the surface of a-SiC:H photoelectrode and it appears as if something deposited on it as shown the insert magnified image in Fig.13(b), while before the test the surface is smooth (insert magnified image in Fig.13.(a)). More work is needed to understand this.

We have also noted changes, in the J-V characteristic after the 100-hour test, a-SiC:H photoelectrode when exposed to air as shown in Fig.14. We note that after 10-day exposure to air, J-V characteristic remains unchanged. After exposing to air for 60 days J_{ph} decreases and photocurrent onset shifts cathodically. The reason probably is that extended exposure to air could leads to SiO_x formation on the surface of a-SiC:H photoelectrodes and its thickness could be time dependent.

4.2 Effect of SiO_x on the surface of a-SiC:H

As seen in Fig.12 and 14, the J_{ph} and its onset of a-SiC:H photoelectrodes could vary under different conditions. As discussed above, there is a possibility of SiOx layer formation on the surface of a-SiC:H photoelectrodes. Using X-ray photoelectron spectroscopy (XPS), we have investigated the surface of a-SiC:H films. Fig.15 shows the XPS spectra for an a-SiC:H film conditions of "as-is" and after etching with hydrofluoric acid (HF concentration of 48%) for different etching times, 10 to 60 seconds. It is clearly seen that there is a very thin SiO_x (a few nm thick) which exists on the a-SiC:H surface, as evident by the peak around 104eV which is associated with the Si-O bonds in SiO_x. The SiO_x layer becomes thinner as the HF etching time become longer, and disappears eventually after an HF dip for 30 seconds. The peak around 101eV related to Si peak, remains the same.

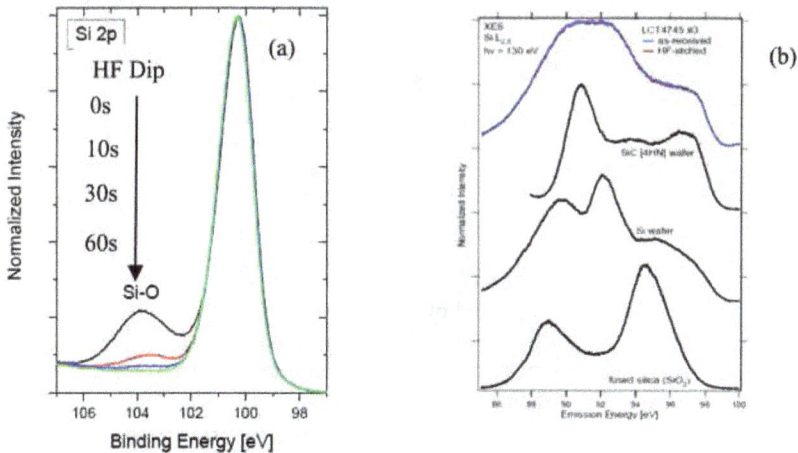

Fig. 15. Changes in XPS (a) and XES (b) curves of a-SiC:H films with HF etching of surface.

The XES curves of before and after HF dip are completely superimposed, which suggests that HF dip does not change the composition of a-SiC:H, as shown in Fig. 15(b). XES curves of crystal SiC wafer, Si wafer and SiO_2 are also shown in Fig. 15 (b) as well as that of a-SiC:H. Crystal SiC wafer has characteristic peaks at 91eV and 98eV while crystal Si wafer exhibits peaks at 90eV and 92eV. One can see that the data for a-SiC:H curve includes crystal Si and SiC characteristic peaks but not SiO peaks which would be at 87eV and 95eV a-SiC:H.

This data suggests that SiO_x grows on the surface of a-SiC:H after deposition and when exposed to air and not during the fabrication process.

In order to investigate the effect of thin SiO_x on the J-V characteristic of the a-SiC:H photoelectrodes, the thickness of SiO_x layer was systematically reduced using HF dip. Before the test, a-SiC:H photoelectrodes were measured, then dipped in HF for 10 to 30 seconds, and measured again. Fig.16 shows a comparison of J-V characteristics before and after the HF dip. It is seen that after HF dip, the J_{ph} increases noticeably (absolute value), e.g., from -2.92mA/cm² to -5.54mA/cm² for HF-dip for 10-second, and -6.3mA/cm² for HF-dip for 30-second at -1.4V vs.Ag/AgCl. Meanwhile, the photocurrent onset shifts anodically by about 0.23V for HF-dip for 30 seconds. Further increasing the dip time beyond 30 seconds, corrosion of a-SiC:H film was clearly evident (as seen clearly by naked eyes). Interestingly, after the a-SiC:H photoelectrode was removed from the electrolyte and exposed to air for 1.5 hours, the J-V characteristics was the same as after the HF dip (not shown here); however, after exposing to air for 67 hours, J_{ph} degraded and completely returned to its initial value as shown in Fig.16 (red one). These results confirm without any doubt that the thin SiO_x layer on the surface of a-SiC:H photoelectrodes indeed affects both J_{ph} and its onset.

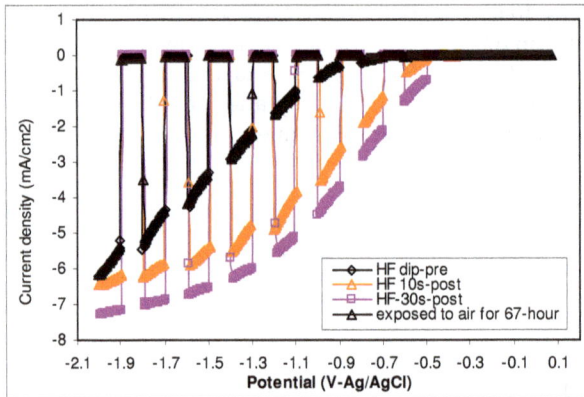

Fig. 16. J-V characteristics of a-SiC:H photoelectrode before and after HF-dip

Same HF dip experiments were repeated by the group in Hawaii (HNEI), in which SCE was used as RE and Xenon lamp calibrated to AM1.5 as light source. The results are the same as shown above in Fig.16, except that higher J_{ph}, (> 8mA/cm²) was observed and possibly due to the use of a different light source.

Comparing Fig.16 with Fig.14, though the effect of SiO_x on J-V characteristics of a-SiC:H photoelectrodes is verified to some extent, J_{ph} and its onset variations are significantly different. J-V characteristic of a-SiC:H photoelectrode after 100-hour durability test and exposed to air for 60 days returns, but not comes back its original value, while after HF-dip and exposure to air only for 67 hours J-V characteristic reverts to its initial value. Apparently, the durability test has modified the surface of a-SiC:H photoelectrodes, resulting in a favorable interface which facilitates photocurrent generation. More work is underway to understand the surface variation during durability test and its effect on the J_{ph} and its onset.

4.3 Integration of the a-SiC:H photoelectrode with a-Si tandem device

The above results show that the a-SiC:H photoelectrode exhibits high photocurrent (i.e., up to ~8 mA/cm^2), and good durability in a pH2 electrolyte for up to ~100 hours. Its main drawback, however, is the non-ideal surface band structure. Our theoretical analysis showed that the hydrogen evolution reaction is thermodynamically allowed at the surface of the a-SiC:H photoelectrode, since the photogenerated electrons are of energy which is higher than the redox potential of H$^+$/H$_2$ half-reaction. On the other hand, to promote oxygen evolution at the counter electrode, a minimum external bias of ~ -1.4 V is needed to bring the quasi-Fermi energy level of photogenerated holes below H$_2$O/O$_2$ redox potential (Hu, et al., 2008).

In order to solve this non-ideal valence band edge alignment problem, an a-Si:H tandem solar cell was integrated into the PEC cell to form a hybrid PV/a-SiC:H configuration, as shown in Figure 17(a). The substrate used for the hybrid PEC device was typically Asahi U-type fluorine-doped SnO$_2$ coated glass. Other types of substrates such as stainless steel and ZnO coated glass were also used for comparison purposes. The a-Si:H tandem solar cell which was used in the hybrid PEC device when fabricated into a solid state device exhibited Voc = 1.71 V, Jsc=7.02 mA/cm^2, FF=0.74, and efficiency of ~9%, as shown in Figure 17(b).

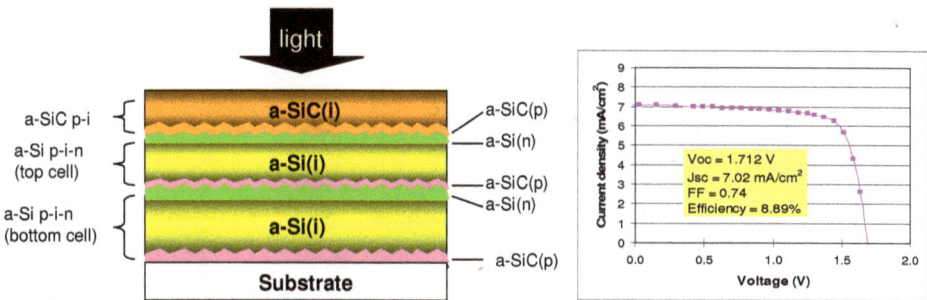

Fig. 17. (a) Configuration of the hybrid PEC device and (b) J-V curve of a-Si tandem device

4.4. Flat-band potential

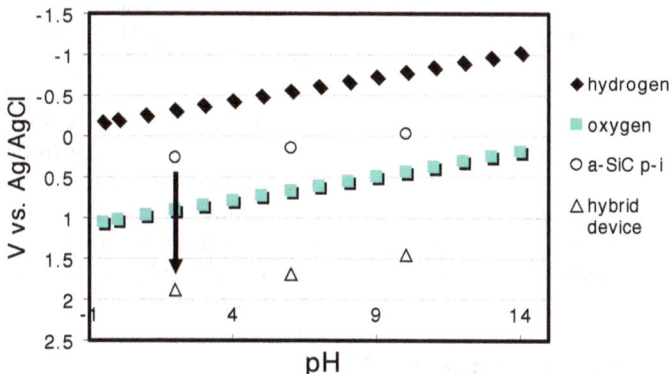

Fig. 18. V$_{fb}$ vs. pH for the a-SiC:H photoelectrode and the hybrid PEC device

Fig.18 shows the flat-band potential (V_{fb}), determined by the illuminated open-circuit potential (OCP) method, as a function of pH of electrolyte for both the a-SiC:H photoelectrode (open circles) and the hybrid PEC device (open triangles). The change of the V_{fb} with pH nearly exhibits a slope of ~60 mV/pH, as predicted by the Nernst equation (Memming, 2000). We note in Fig. 18, that at pH2, V_{fb} =+0.26 V (vs. Ag/AgCl) for the a-SiC:H photoelectrode, whereas in the case of the hybrid device, the V_{fb} significantly by ~ +1.6 V (as indicated by the red arrow) and is below the H_2O/O_2 half-reaction potential (by +0.97 V) and is in an appropriate position to facilitate water splitting. Figs.19 (a) and (b) show the current density vs. potential characteristics for hybrid PEC device fabricated on different substrates, SnO_2 and ZnO coated glass and SS, and measured in the pH2 buffered electrolyte (sulphamic acid solution with added potassium biphthalate) using the 3-electrode and 2-electrode setup respectively. In the 2-electrode setup, there was no reference electrode and contained only the working electrode (hybrid PEC device) and the counter electrode which was ruthenium oxide (RuO_2) rather than conventional platinum (Pt).

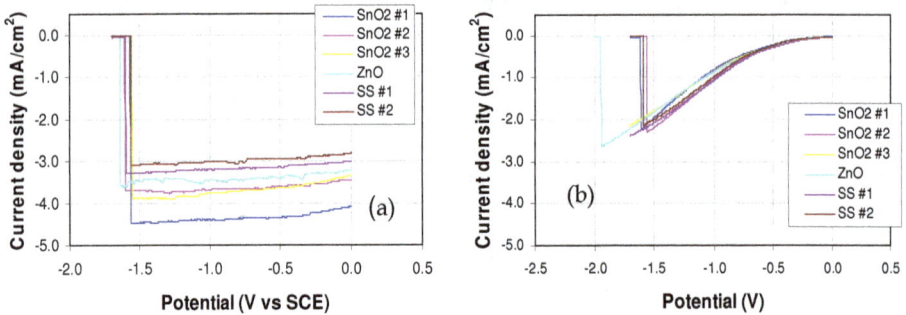

Fig. 19. Current density vs. potential characteristics measured in (a) 3-electrode and (b) 2-electrode setup.

From Fig.19 (a), we see the saturated photocurrent of the hybrid cell using different substrates is in the range of 3-5 mA/cm². The larger photocurrent using SnO_2 (>4 mA/cm²) coated glass substrate is due to the inherent texture of the SnO_2 which enhances internal photon absorption. More significantly, we see that the photocurrent density of ~0.3 mA/cm² occurs at a zero potential using the 2-electrode setup (Fig.19 (b)). Hydrogen production was observed in a short-circuit condition (Hu, et al., 2009).

It should be noted that compared with the 3-electrode case the photocurrent measured in the 2-electrode setup (even using RuO_2 counter electrode) is much lower, suggesting limiting factors. We have noted that the over-potential loss in the 2-electrode setup can be due to, (1) type of electrolyte used, (2) type of counter electrode used and (3) formation of thin SiO_x layer on the a-SiC:H surface. Initial results have shown that, after dipping the hybrid device into 5% hydrofluoric (HF) acid for 30 seconds and using RuO_2 as the counter electrode, the photocurrent is enhanced from 0.33 to 1.33 mA/cm² at zero bias in two-electrode setup.

4.5 Durability of the hybrid PEC device

The test was performed in the pH2 buffered electrolyte, with Pt as the counter electrode. During test, a constant current density of 1.6 mA/cm² was applied to the device, while the voltage (potential) across the sample was recorded over a 148-hour period. The current density vs. potential characteristics of the device prior to and after 22, 48 and 148-hour tests were measured in both the 3- and 2-electrode setups. Throughout these durability tests, H_2 production from the hybrid device occurred. Fig.20 (a) shows the current vs. potential curves measured prior to and after 148-hour test. These results show that the dark current shows almost no change, and hence no corrosion occurs in the hybrid device, after the 148-hour test, as is evident also in Fig.20 (b) and (c).

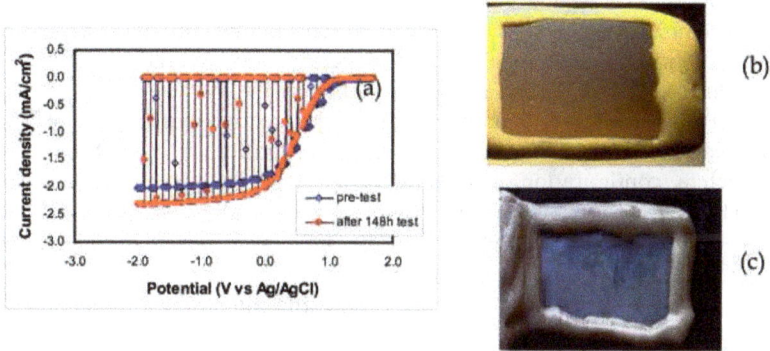

Fig. 20. (a) Current density vs. potential characteristics measured prior to and 148-hour test. (b) Photo images of the hybrid PEC device prior to and (c) after a 148-hour test in pH2 electrolyte.

5. Pathway to 10% STH efficiency in PV/a-SiC:H device

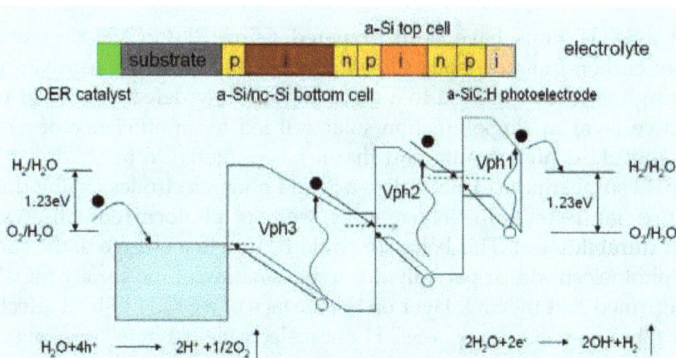

Fig. 21. The schematic energy band diagram of a hybrid PV/PEC cell containing a-SiC:H photoelectrode and a-Si :H/a-Si :H (or nc-Si :H) tandem solar cells

In the previous sections, we have shown that a-SiC:H photoelectrodes are durable in electrolyte for up to 100 hours (so far tested). The J_{ph}, measured in a three-electrode setup, could be larger than 6mA/cm² (at -1.4V vs. Ag/AgCl) after removing SiO_x layer on the surface. The extra bias is needed because, (1) the non-ideal energy band edge alignment to the H_2O/O_2 redox potential, as confirmed by the flatband voltage measurement and (2) various over-potential losses, e.g., due to the surface SiO_x barrier layer and various interfaces. Incorporating a-Si tandem photovoltaic device could eliminate the external bias as suggested by previous data. The schematic energy band diagram for a PV/a-SiC:H cell is shown in Fig.21. Since a-Si:H tandem solar cell could provide a photovoltage ($V_{ph2}+V_{ph3}$) of >1.8V (Shah, et al., 2004), while the a-SiC:H photoelectrode could provide a photovoltage (V_{ph1}) >0.5V (as shown by the initial tests), hence the total photovoltage in such a configuration (sum of $V_{ph1} + V_{ph2} + V_{ph3}$) is expected to be >2.3 V. Thus, such a PV/a-SiC:H structure can provide needed voltage for water splitting, as confirmed by the results described in the previous sections.

Further enhancement in the STH efficiency can be achieved by employing a nano-crystalline silicon single junction solar cell (nc-Si:H) in the tandem solar cell, and integrated with the a-SiC:H photoelectrode as described in Fig.21. Although the Voc of nc-Si single junction is ~0.55V (Mai, et al., 2005), lower than that in a-Si single junction, but the sum of $V_{ph1} + V_{ph2} + V_{ph3}$ is expected to be higher than 1.5V, still large enough for water splitting. It is possible that with such a configuration, a J_{ph}>8mA/cm^2 could be generated, leading to STH efficiency up to 10% as shown in table 1.

PEC (a-SiC:H)		PV cell		STH (%)
Eg	J_{ph} (mA/cm2)	configuration	J_{ph} after filtered by a-SiC:H (mA/cm^2)	
2	8.85	a-Si/a-Si	6.7	8.24
2.3	8.55	nc-Si/a-Si	8.85	10.52

Table 1. The PV/a-SiC:H structure PEC simulation results

6. Conclusion

State-of-the-art a-SiC:H films have been prepared using RF-PECVD deposition technique. Incorporation of carbon in amorphous silicon network increases the bandgap to >2.0eV and adding H$_2$ during fabrication has led to a material with low defects. A-SiC:H with Eg=2.0eV used as the active layer in single junction solar cell led to an efficiency of ~7%, which also indicated that a-SiC:H is high-quality and that it has potential to be used as photoelectrode. Immersing in pH2 sulphamic acid electrolyte a-SiC:H photoelectrodes exhibit durability for up to 100 hours (so far tested). J_{ph} increases as well as photocurrent onset shifts towards anodically after durability test. This behavior could be due to a change in the surface structure of the a-SiC:H photoelectrode, or partially due to elimination of the surface SiO$_x$ layer. HF etch experiment confirmed that the SiO$_x$ layer on the surface of a-SiC:H indeed affects both J_{ph} and its onset. After removing SiO$_x$ layer, a-SiC:H photoelectrode exhibited a J_{ph} over 6 mA/cm^2 at potential -1.4V (vs. Ag/AgCl), compare to that less than 4mA/cm^2 (vs. Ag/AgCl). More analysis needs to be done to understand the mechanisms and improve the interface between a-SiC:H and electrolyte and hence to increase J_{ph}. Our initial PV/a-SiC:H used as photoelectrode has exhibited ~1.33mA/cm^2 current density under zero volt external bias, and the photocurrent onset shifts enormously, from ~ -0.6V to ~ +1.2V, or by a net ~1.6v. Hydrogen production has been demonstrated in this type of hybrid PEC cell. It exhibits good durability in aqueous electrolytes for up to ~150 hours. Work on further increasing the photocurrent in such a PV/a-SiC:H device is underway. We have also shown by simulation that it is possible to achieve STH efficiency >10% using such PV/a-SiC:H devices.

7. Acknowledgments

The work is supported by US Department of Energy under contract number DE-FC36-07GO17105. The authors would like to thank Ed Valentich for his assistance in sample fabrication.

8. Reference

Bak, T.; Nowotny, J.; Rekas, M.; Sorrell, C. C. (2002). Photo-electrochemical hydrogen generation from water using solar energy. Materials-related aspects, *international Journal of hydrogen energy*, 27 (October 2002) 991-1022, ISSN: 0360-3199

Cody, G.D.; Tiedje, T.; Abeles, B.; Moustakas, T.D.; Brooks, B.; and Goldstein, Y.; (1981); Disorder and the optical absorption edge of hydrogenated amorphous silicon, *Journal De Physique* colloque C4, (October 1981) C4-301-304, ISSN:1155-4304

Desalvo, A.; Giorgis, F.; Pirri, C.F.; Tresso, E.; Rava, P.; Galloni, R.; Rizzoli, R.; and Summonte, C.; (1997), Optoelectronic properties, structure and composition of a-SiC:H films grown in undiluted and H_2 diluted silane-methane plasma, *Journal Applied Physics* 81 (June 1997) 7973-7980, ISSN: 0021-8979

Deutsch, T.G. ; Head, J. L.; Turner, J. A.; (2008). Photoelectrochemical Characterization and Durability Analysis of GaInPN Epilayers, *Journal of Electrochemcal Society*, Vol. 155, (July 2008), B903-B907, ISSN: 0013-4651Fujishima, Akira; & Honda, Kenichi; (1972), Electrochemical Photolysis of Water at a Semiconductor Electrode, *Nature*, 238 (July 1972) 37-38, ISSN: 0028-0836

Gerischer, H. & Mindt, W. (1968).The Mechanisms of the Decomposition of Semiconductors by Electrochemical Oxidation and Reduction, Electochimica Acta, 13 (June 1968) 1329-1341, ISSN: 0013-4686

Gratzel, M. (2001). Photoelectrochemical cells, *Nautre* Vol.414 (November 2001) 338-344, ISSN: 0028-0836

Hassan, A. & Hartmut, S. (1998). The German-Saudi HYSOLAR program, *International Journal of hydrogen energy* 23 (June 1998) 445-449, ISSN: 0360-3199

Helmut, T. (2008). Photovoltaic hydrogen generation, *International Journal of Hydrogen energy*, 33 (November 2008) 5911-5930, ISSN: 0360-3199

Hu, Z. H.; Liao,X. B.; Diao, H. W.; Kong, G.L.; Zeng, X. B.; and Xu, Y.Y.; (2004), phous silicon carbide films prepared by H2 diluted silane–methane plasma, *Journal of Crystal Growth*, Vol. 264 (March 2004) 7-12, ISSN: 0022-0248

Hu, J.; Zhu, F.; Matulionis, I.; Kunrath, A.; Deutsch, T.; Miller, L. E.; Madan, A.; (2008), Solar-to-Hydrogen Photovoltaic/Photoelectrochemical Devices Using Amorphous Silicon Carbide as the Photoelectrode, *Proc. 23rd European Photovoltaic Solar Energy Conference, Feria Valencia, 1-5 September 2008*, #1AO.5.6

Hu, J.; Zhu, F.; Matulionis, I.; Deutsch, T.; Gaillard, N.; Miller, L. E.; Madan, A.; (2009), Development of a hybrid photoelectrochemical (PEC) device with amorphous silicon carbide as the photoelectrode for water splitting, *Symposium S: Materials in Photocatalysis and Photoelectrochemistry for Environmental Applications and H2 Generation , MRS Spring Meeting, San Francisco, April 13-17, 2009*, # 1171-S03-05

Hollingsworth, R.; Bhat, P.; and Madan, P.; (1987); Amorphous silicon carbide solar cells, *IEEE Photovoltaic Specialists Conference, 19th*, New Orleans, LA, May 4-8, 1987, Proceedings (A88-34226 13-44) 684-688

Khaselev, O. & Turner, J. A.; (1998), A Monolithic Photovoltaic-Photoelectrochemical Device for Hydrogen Production via Water Splitting, *Science* Vol.280, (April 1998), 425-427, ISSN: 0036-8075

Kuznetsov, A.M. & Ulstrup, J.; (2000). Theory of electron transfer at electrified interfaces, *Electrochimica Acat* 45 (May 2000) 2339-2361, ISSN: 0013-4686

Lichta, O. S.; Wang, B. ; Mukerji, S. ; Soga, T.; (2001). Over 18% solar energy conversion to generation of hydrogen fuel; theory and experiment for efficient solar water splitting, *International Journal of Hydrogen Energy* 26(July 2001), 653–659, ISSN: 0360-3199

Madan, A. & Shaw, M. P., (1988), The Physics and Applications of Amorphous Semiconductors, *Academic* Press, ISBN-13: 9780124649606

Mai, Y.; Klein, S.; Carius, R.; Stiebig, H.; Geng, X.; (2005); Open circuit voltage improvement of high-deposition-rate microcrystalline silicon solar cells by hot wire interface layers, *Appllied Physics Letters 87* (2005) 073503, ISSN:0003-6951

Mary, D.A. & Arthur, J.N. (2008). *Nanostrustrued and photoelectrochemical Systems for Solar Photo Conversion*, Imperial College Press, ISBN-13: 978-1-86094-255-6

Masayoshi, U.; Atsuko, K.; Mihoko, O.; Kentarou, H.; and Shinsuke, Y., (2005). Photoelectrochemical study of lanthanide titanium oxides, Ln2Ti2O7 (Ln = La, Sm, and Gd), *Journal of Alloys and Compounds*, 400, (September 2005) 270-275, ISSN: 0925-8388

Mathews, N.R.; Miller, Eric, L.; Sebastian, P.J.; Hernandez, M.M.; Mathew, X.; and Gamboa, S.A.; (2004), Electrochemical characterization of a-SiC in different electrolytes, *International Journal of Hydrogen Energy*, 29 (August 2004) 941-944, ISSN: 0360-3199

Matulionis, I.; Zhu, F.; Hu, J.; Deutsch, T.; Kunrath, A.; Miller, L. E.; Marsen, B.; Madan,A; (2008).Development of a corrosion-resistant amorphous silicon carbide photoelectrode for solar-to-hydrogen photovoltaic/photoelectrochemical devices, *proceedings of the SPIE Conference on Solar Hydrogen and Nanotechnology*, Vol. 7044, 7044-11, ISBN: 9780819472649, San Diego, CA, USA, August 11-15, 2008, San Diego

Memming, Rudiger. (2000). *Semiconductor Electrochemistry*, Wiley-VCH, ISBN-13: 978-3527301478

Miller, E. L. & Rocheleau, E. R. (2001), Photoelectrochmical hydrogen production, *Proceedings of the 2001 DOE Hydrogen Program Review NREL/CP-570-3053*

Miller, E. L.& Rocheleau, E. R. (2002). Photoelectrochemical production of hydrogen, Proc. 2002 US DOE Hydrogen Program Review, NREL/CP-610-32405

Miller, E. L. & Rocheleau, E. R.; Deng, X.; (2003). Design considerations for a hybrid amorphous silicon/photoelectrochemical multijunction cell for hydrogen production, International Journal of Hydrogen Energy, 28 (June 2003) 615-623, ISSN: 0360-3199

Morisaki, H.; Watanabe, T.; Iwase, M.; and Yazawa,; (1976), Photoelectrolysis of water with TiO2-covered solar-cell electrodes, *Applied Physics Letter*, Vol, 29, (September1976) 338-340, ISSN: 0003-6951

Murphy, A.B.; Barnes, P.R.F.; Randeniya, L.K.; Plumb I.C.; Grey I.E.; Horne M.D.; and Glasscock, J.A.; (2006). Efficiency of solar water splitting using semiconductor electrodes, *International Journal of Hydrogen energy* 31 (2006) 1999-02017, ISSN: 0360-3199

Narayanan, R. & Viswanathan, B. (1998). *Chemical and Electrochemical Energy Systems*, University Press (India) Limited, ISBN: 8173710694

Nelson, A. K. & Thomas, L. G., (2005). Design and characterization of a robust photoelectrochemical device to generate hydrogen using solar water splitting, *International Journal of hydrogen energy* 31 (September 2006) 1658-1673, ISSN: 0360-3199

Nelson, A. K.& Thomas L.G.; (2008). Design and characterization of a robust photoelectrochemical device to generate hydrogen using solar water splitting, *International Journal of Hydrogen Energy 31* (November 2008) 1658-1673, ISSN: 0360-3199

Nozik, A. J.; (1975), Photoelectrolysis of water using semiconducting TiO2 crystals, *Nature*, 257 (October 1975) 383-385, ISSN: 0028-0836

Richard, R.E.; Eric, L. M.; Anupam, M.; (1998). High-effciency photoelectrochemical hydrogen production using multijunction amorphous silicon photoelectrodes, *Energy & Fuels*, 12 (January 1998); 12:3–10, ISSN: 0887-0624

Shah, A.V; Schade, H.; Vanecek, M.; (2004); Thin-film silicon solar cell technology, *Progress in Photovoltaics: Research and Applications*, 12 (March 2004) 113-142, ISSN: 1062-7995

Simonds, B. J. (a); Zhu, F. Zhu; Gallon, J.; Hu, J.; Madan, A.; and Taylor, C.; (2009); Defects in Hydrogenated Amorphous Silicon Carbide Alloys using Electron Spin Resonance and Photothermal Deflection Spectroscopy, *Mater. Res. Soc. Sym. Proc. Vol.1153, April 13-17, 2009, San Francisco, CA*, paper # 1153-A18-05

Simonds, B. J. (b); Zhu, F. Zhu; Hu, J.; Madan, A.; and Taylor, C.; (2009); Defects in amorphous silicon carbide and their relation to solar cell device performance, *proceedings of the SPIE Conference on Solar Hydrogen and Nanotechnology*, Vol. 7409, 7408-3, ISBN: 9780819476999, San Diego, CA, USA, August 2009, San Diego

Sebastian, P. J.; Mathews, N. R.; Mathew, X.; Pattabi, M.; and Turner, J.; (2001), Photoelectrochemical characterization of SiC, *International Jounal of Hydrogen Energy* 26 (February 2001) 123-125, ISSN: 0360-3199

Solomon, I.;(2001); Amorphous silicon–carbon alloys: a promising but complex and very diversified series of materials, *Applied Surface Science* 184 (December 2001) 3-7, ISSN: 0169-4332

Srivastava, O.N.; Karn, R.K.; and Misra, M.; (2000), Semiconductor-septum photoelectrochemical solar cell for hydrogen production, *International Journal of Hydrogen Energy*, 25 (June 2000) 495-503, ISSN: 0360-3199

Stavrides, A.; Kunrath, A; Hu, J.; Matulionis, I.; Marsen, B.; Cole, b.; Miller, E.; and Madan, A.; (2006), Use of amorphous silicon tandem junction solar cells for hydrogen production in a photoelectrochemical cell, *proceedings of the SPIE Conference on Solar Hydrogen and Nanotechnology*, Vol. 6340, 63400K, ISBN: 9780819464194, San Diego, CA, USA, August 2006, SPIE, San Diego

Tamaura, Y.; Steinfeld, A.; Kuhn, P.; and Ehrensberger, P. (1995). Production of Solar Hydrogen by a Novel, 2-step, Water-splitting Thermochemical cycle, *Energy* Vol. 20, No.4 (1995) 325-330, ISSN: 0360-5442

Tokio O. (1979). *Solar-Hydrogen Energy Systems*, Pergamon Press (Oxford, New York), ISBN: 0080227139

Yamada, Y.; Matsuki, N.; Ohmori, T.; Mametsuka, H.; (2003). One chip photovoltaic water electrolysis device, *International Journal of Hydrogen Energy* 28 (November, 2003) 1167– 1169, ISSN: 0360-3199

Yoon, D. H.; Suh, S. J.; Kim, Y. T.; Hong, B.; Jang, G. E.; (2003); Influence of Hydrogen on a-

SiC:H Films Deposisted by RF PECVD and Annealing Effect, *Journal of Korean Physic Society* Vol. 42 p 943-946, ISSN 0374-4884

Zhu, F.; Hu, J.; Matulionis, I.; Deutsh,T.; Gailard, N.; Kunrath, A.; Miller, L.E.; Madan, A.; (2009), Amorphous silicon carbide photoelectrode for hydrogen production directly from water using sunlight, *Philosophical Magazine* 89:28 (October 2009) 2723-2739, ISBN: 1478-6443

4

Organic Solar Cells Performances Improvement Induced by Interface Buffer Layers

J. C. Bernède[1], A. Godoy[2], L. Cattin[1], F. R. Diaz[3],
M. Morsli[1] and M. A. del Valle[3]

[1]*Université de Nantes, Nantes Atlantique Universités, LAMP, EA 3825, Faculté des Sciences et des Techniques, 2 rue de la Houssinière, BP 92208, Nantes, F-44000*
[2]*Facultad Ciencias de la Salud, Universidad Diego Portales. Ejército 141. Santiago de Chile*
[3]*Facultatd de Quimica, PUCC, Casilla 306, Correo 22, Santiago,*
[1]*France*
[2,3]*Chile*

1. Introduction

The energy sector has a constrained future, since increasing demand coincides with "prise de conscience" of the negative implications of fossil energy use. Global warming is finally a clear evidence of the fundamental idea of the "old" Newtonian physics: there is no action without reaction. Fundamental principle neglected by the occidental world during the last century. That is to say, we cannot continue to emit continuously carbon dioxide, nitrogen dioxide… and others pollutants produced from the burning of fossil energies into our environment without suffering the consequences. Some environmental scientists have highlighted this problem for some time [Lüthi et al., *Nature*, 2008], but only now are some governments giving the issue the attention that it deserves. Man-made climate change is one of the greatest threats our world faces. Renewable energies issued from our natural environment, such as wind power, solar thermal, photovoltaic, geothermal heat, marine and hydro power…, can help reduce our dependence on fossil energies. The present review is dedicated to photovoltaic energy and more precisely to some specific photovoltaic devices based on organic materials.

Photovoltaic cells belong to the family of the optoelectronic devices. As evidenced by their denomination, such devices use the optical and electronic transport properties of different materials to either produce electromagnetic radiation (light emitting diodes) or to generate electricity (photovoltaic cells -PV cells). Photovoltaic cells also called solar cells are used to generate electrical power. A PV cell is a device based on the photoconductive properties of semiconductor materials -for carriers generation- coupled with the ability of these semiconductors to form junctions -for carriers separation. The photoconductivity is the process in which electromagnetic energy is absorbed by a material and converted to excitation energy of electric charge carriers so that the material becomes quite conductor. When irradiated by a light, PV cells produce electrical energy across any connected external load. When irradiated without load a PV cell produces a maximum photogenerated voltage V_{oc}, the open-circuit voltage. When shorted, the PV cell produces the maximum short circuit current I_{sc}. When connected to a load the power output of the cell is given by the voltage current product VxI. The maximal power generated possible is $V_{oc}xI_{sc}$. In fact the maximum power a PV cell is able generating depends on the dark I-V characteristics, that is to say on the diode properties of the junction constituting the device. When the load value is optimised, the maximum power provided by the cell is $Pm = V_m x I_m$. A figure of merit called the fill factor, FF, for the PV cells is given by:

$$FF = V_m x I_m / V_{oc} x I_{sc} \ (1).$$

Up to now, inorganic materials are used in photovoltaic cells. Crystalline, polycrystalline and amorphous silicon represent more than 95 % of the world production, while CdTe and $Cu(In,Ga)Se_2$ (CIGS) are now emerging in the market. Crystalline (or polycrystalline) devices allow achieving efficiencies up to 25%. However, efficient crystalline (or polycrystalline) devices are difficult and expensive to produce and the pay-back time of such modules is around three years. Traditionally, optoelectronic devices were grown using inorganic compounds. However, some years ago, research devoted to organic light emitting diodes (OLEDs) encounter an unexpected success [Jain et al., *Semiconductors and semimetals*, 2007] and they are now available on the market. Moreover it has been shown that the quantum efficiency of the electron transfer from an excited polymer to fullerene (C_{60}) is very high [Xiong Gong et al. *Sciences*, 1992]. So, since the pioneering work of Tang [Tang, *Appl. Phys. Lett.*, 1986] the interest devoted to organic solar cells has been raising very fast, which has undergone a gradual evolution of the energy conversion efficiency, η, from less than 1% to more than 5% [Kim et al., *Sciences*, 2007, Xue et al., *J. Appl. Phys.*, 2005]. These significant progresses demonstrate that organic solar cells are a potential avenue to low cost next generation solar cells. However, some efforts are still necessary to improve the cell efficiency and lifetime. To overcome the quite narrow absorbance domain of the organic photoactive layer, several approaches such as low band gap organic material, incorporation of metal nanostructures, use of inorganic optical spacer between the active layer and the electrode can be used. It is also well known that carriers exchange at interfaces organic material/electrode can greatly influence device performance. In the present review, based on our recent studies, we will discuss more specifically possible device improvement through interface optimisation. The plan of the manuscript is as follow, after recalling some generality on organic solar cells and the classical interface theory in semiconductors, impact of electrode/organic interface properties on cells performances will be discussed using different published results, and more specifically studied from our last results. All the results will be critically discussed in the context of how to improve the fundamental understanding of interface behavior to enhance solar cells performance.

2. A short comparison with organic light emitting diodes

As said above, the development of efficient organic displays based on organic light emitting devices (OLEDs) has shown that organic electronic components are viable. Those displays are now developed using low cost technology and these new technologies development for OLED can be tested for PV solar cells realisation [Bernède et al., *Current Trends in polymer Sciences*, 2001]. Basically the underlying principle of a photovoltaic solar cell is the reverse of the principle of OLED Figure 1.

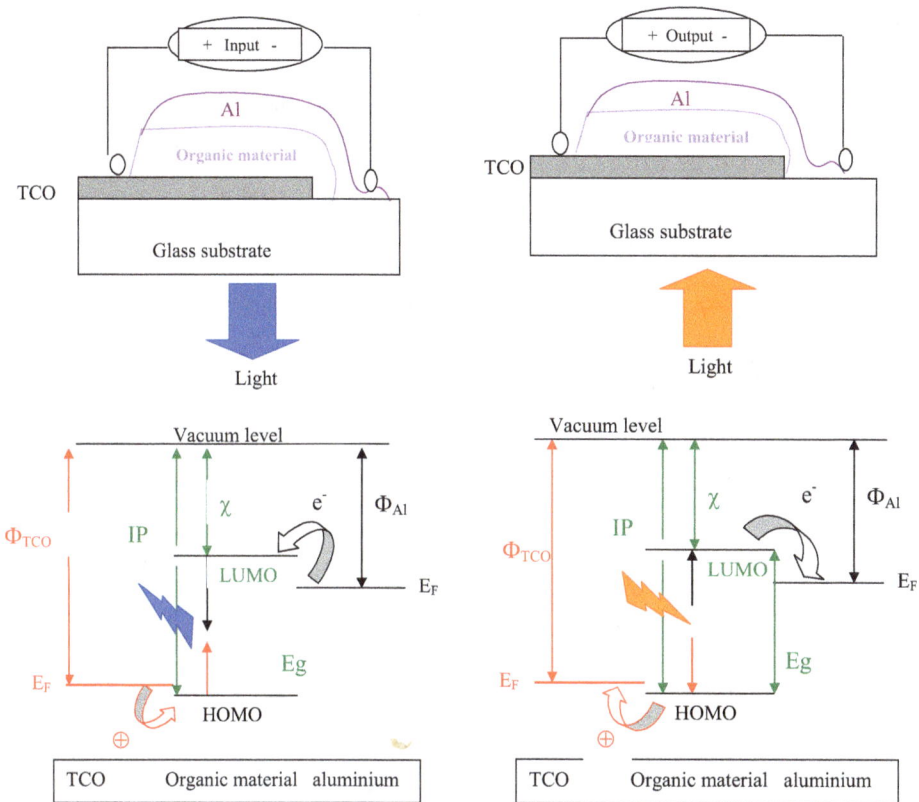

Fig. 1. Principle of an OLED (left) and a solar cell (right) (Band scheme without contact)

In OLED, electrons are injected at the low work function electrode (cathode), while holes are injected at the high work function electrode (anode). At some point in the organic, the electron and hole meet and recombine with light emission. The reverse happens in a PV cell, when light is absorbed an exciton forms. After exciton dissociation, the electron must reach the low work function electrode and the hole the high work function electrode.

In fact, when the organic material is put into contact with electrode, the shape of the band scheme depends on the conductance of the organic material Figure 2.

When the cells are short circuited, the Fermi levels of the electrodes align. If the organic is an insulator, the field profile changes linearly through the cell (fig. 2b). If the organic is a p-type semiconductor a depletion layer forms on the side of the metal with small work function, we have Schottky contact (fig. 2c). Usually　the former scheme is used in OLEDs, the organic films used being quite insulating and the latter scheme is often used in solar cells, the organic active layers being semiconducting.

Almost all organic optoelectronic devices have a planar layered structure, where the organic active layer(s) is (are) sandwiched between two different electrodes. One of them must be transparent. A transparent conductive oxide (TCO) is used, usually indium tin oxide (ITO) because it allows achieving better results. The other electrode is very often aluminium, even if calcium has a better work function, because Al is stable in air while Ca is not. From the above comparison it can be concluded that a device which exhibits high electro-luminescent properties will be a poor solar cell and vice versa. However in both devices families the properties of the contact electrode/organic material are determinant to the efficiency of the devices, and the progress in that field for one device family is very helpful for the other family.

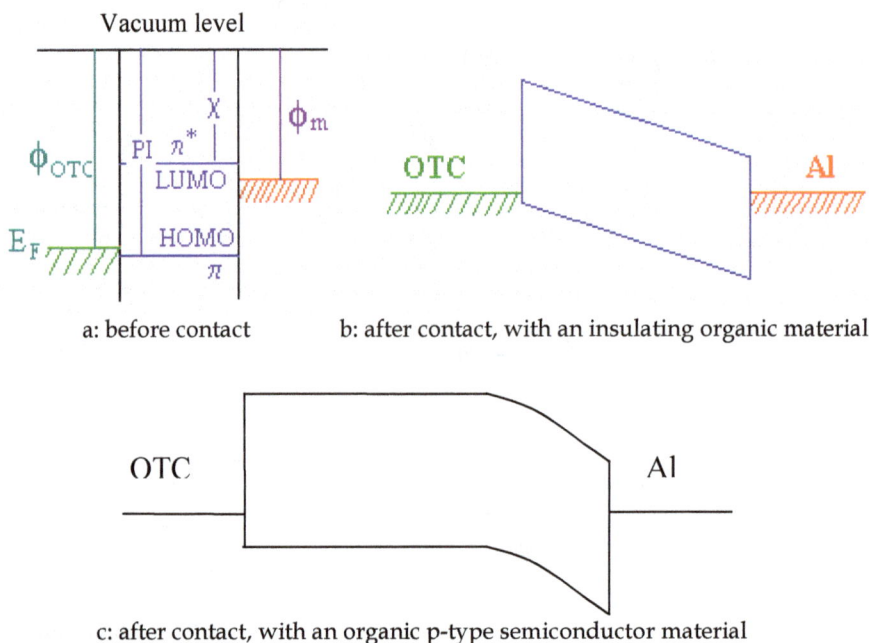

a: before contact   b: after contact, with an insulating organic material

c: after contact, with an organic p-type semiconductor material

Fig. 2. Band scheme of TCO/organic/Al structure

3. Different organic solar cells families

Organic semiconductors, such as macromolecules dyes, dendrimers, oligomers, polymers..., are all based on conjugated π electrons. Conjugated systems are based on an alternation between single and double bonds. The main property related to this conjugation is that π electrons are more mobile than σ electrons. Therefore the π electrons can move by hopping. These π electrons allow light absorption, in the case of solar cells, and emission, in the case of OLEDs. Molecular $\pi-\pi^*$ orbitals correspond respectively to the Highest Occupied Molecular Orbital (HOMO) and Lowest Unoccupied Molecular Orbital (LUMO). For sake of simplicity, such organic material can be regarded as a semiconductor-like material, where the band gap corresponds to the difference between the LUMO and the HOMO.

Photons absorption by inorganic semiconductors produces free electrons and holes, the charge separation is more difficult in organic semiconductors. When a photon of appropriate energy is incident upon organic semiconductor it can be absorbed to produce an excited state called exciton, that is to say an electron-hole pair in a bound state which is transported as a quasi-particle. In organic materials excitons are strongly bounded as a consequence of their low dielectric constant. Organic solar cells belong to the class of photovoltaic cells known as excitonic solar cells [Thompson, Fréchet, *Angnew. Chem. Int. Ed.*, 2008]. The excitons can have appreciable life-time before recombination. To produce photocurrent the electron-hole pair of the exciton must be separated. If not, they can recombine either radiatively (luminescence is a loss mechanism in photovoltaic cells) or non-radiatively with heat production. Therefore after light absorption and exciton formation, the carriers should be separated. Even if not well understand the dissociation occurs at defects, impurities, contacts or any other inhomogeneities. The separation occurs in the electric field induced around the inhomogeneity. If the ionisation takes place at a random defect in a region without an overall electric field, the generated carriers will be lost. To avoid such loss, exciton dissociation should occur in high electric field region associated with a contact or a junction. To produce an internal electric field which occupy a substantial volume of the device, the usual method is to juxtapose two materials with different appropriate properties. One of these materials is an electron donor and the other one is called electron acceptor. The interface between these two materials is called heterojunction. Therefore it is clear that the active donor-acceptor pair governs the separation mechanism. While in the case of inorganic materials the both materials of the heterojunction are clearly identified, the electron acceptor is the n-type material related to its electron excess and the electron donor is the p-type material related to its hole excess, it is not so simple in the case of organic materials. The donor or acceptor nature of an organic semiconductor depends on its carrier mobility which is determined by intrinsic properties of this material. Moreover, it is known that the donor or acceptor character of a material in an organic couple depends also on their relative HOMO and LUMO values. For instance, CuPc, which is a usually an electron donor, has been also used as electron acceptor with a triphenylamine derivative used as donor [Chen et al., *Sol. Energy Mater. Sol. Cells*, 2006]. Therefore an organic material with intermediary HOMO and LUMO values can be used as an electron donor for one organic material and as an electron acceptor for another organic material. In figure 3 we can

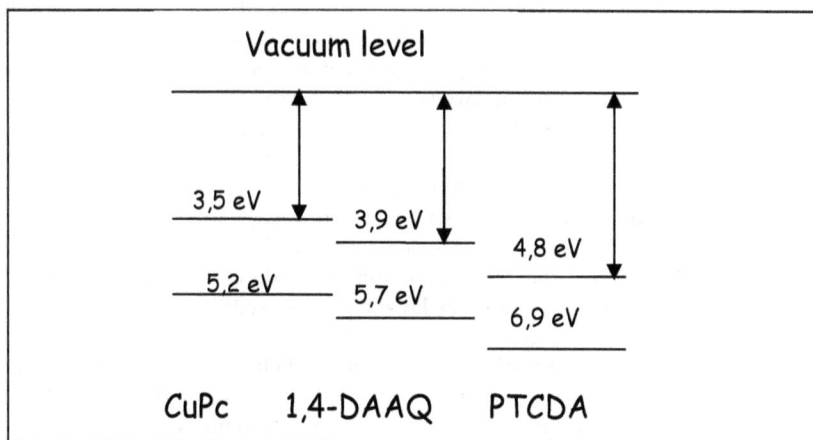

Fig. 3. Relative position of the HOMO and LUMO of CuPc/1,4-DAAQ/PTCDA.

see the example of the 1,4-diaminoanthraquinone (1,4-DAAQ), which is an acceptor relatively to the copper phthalocyanine (CuPc) and a donor relatively to the   perylene 3,4,9,10-tetracarboxylic dianhydride (PTCDA) [Berredjem et al., *Dyes and Pigments,* 2008].

It is well known that exciton dissociation is efficient at the interface between materials with different electron affinity EA (i.e. LUMO) and ionisation potential IP (i.e. HOMO) (Figure 4). The difference in electron affinities creates a driving force at the interface between the two materials that is strong enough to separate charge carriers of photogenerated excitons. EA and IP of the electron acceptor should be higher than those of the donor.

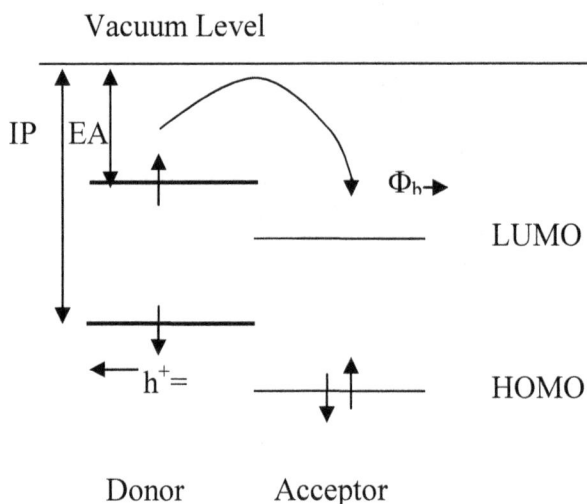

Fig. 4. Charge separation of an exciton into a free electron/hole pair at a donor acceptor interface.

Then the separated electron and hole should be collected at the cathode and anode electrodes respectively. In a crystalline inorganic semiconductor the LUMOs and HOMOs form the conduction and valence band respectively, and carriers transport within a band. It is usually different in the case of the organic semiconductors. In that case intermolecular forces are Van der Waals forces and no large conduction band and valence band are formed. Therefore the charge transport proceeds by hopping between localized states. This means that charge carrier mobility in organic materials is smaller than that in inorganic semiconductors. Mobility is improved with the molecular ordering in the films, it is reduced by impurities as well as by oxygen traps. Therefore charge transport is improved by improving order, purification and no oxygen contamination.

So, the operations of an organic solar cell can be summarized in five fundamental steps:
1. Photon absorption by the active organic layer to create excitons.
2. Diffusion of the excitons to a donor/acceptor interface.
3. Excitons dissociation at the interface and formation of electron-hole carrier pairs.
4. These free carriers should move toward the cathode (electrons) and the anode (holes).
5. The electrons (holes) should be extracted to the external circuit by the cathode (anode).

Up to 1986 very low power conversion efficiencies were achieved (#10^{-6}%) because of low concentration and mobility of free charge carriers. In 1986 Tang reported on a bilayer heterojunction solar cell based on the copper-phthalocyanine (CuPc)/perylene tetracarboxylic derivative with 1% energy conversion efficiency [Tang, *Appl. Phys. Lett.*, 1986]. In spite of numerous efforts to improve the cells performances, there was no breakthrough and it remained around 1%. However, during the last ten years a new and strong interest has been devoted to organic photovoltaic cells, motivated by two recent developments in the organic semi-conductor field. First it has been shown that the quantum efficiency of the electron transfer from an excited polymer to C_{60} is very high and the transfer is very fast, [Sariciftci et al., *Sciences*, 1992; Sariciftci et al., *Appl. Phys. Lett.*, 1993] , which is promising for change carrier separation in PV cells. Secondly the development of efficient organic displays based on organic light emitting devices (OLEDs) has shown that organic electronic components are viable. Those displays are now developed using low cost technology and these new technologies development for OLED can be tested for PV solar cells realisation [Bernède et al., *Current Trends in polymer Sciences*, 2001]. Therefore OLED technology allows drawing a research guideline in organic PV cells. Since the beginning of the 21st century, power conversion efficiency is increasing rapidly and the organic photovoltaic devices are getting closer to commercialisation [Kim, Lee, Chin, *Sol. Energy Mater. Sol. Cells.*, 2009].

Two technological routes are mainly used: the deposition of polymers and nanoparticles from solution by spin coating technique and the vacuum sublimation of small molecules. The former route has given the concept of bulk heterojunction solar cells (BHJ), while the later gives the multi-heterojunction solar cells family. Development of BHJ interpenetrating network structures using polymer blended with a fullerene derivative allows overcoming inherent disadvantageous very short exciton diffusion length, which improves significantly cells performances. Controlled morphology of poly-paraphenylene-vinylene/fullerene blend derivatives increased conversion efficiency up to 2.5% in 2001 [Shaheen et al., *Appl. Phys. Lett.*, 2001]. A post-annealing treatment improved efficiency to 3.5 [Padinger et al., *Adv. Funct. Mater.*, 2003]. To day, the best energy conversion efficiencies (>6%) have been reach with poly-3(hexyl-thiophene) (P3HT) serving as electron donor and [6,6]-phenyl-C_{61}-butyric acid methyl ester (PCBM), which is a fullerene derivative as electron acceptor [Kim et al., *Science*, 2007].

In the multi-heterojunction solar cell family the simplest efficient configuration is based on a bilayer system between two electrodes. About these electrodes, one should be transparent and the other highly reflexive. The bilayer is composed of, at least, one absorbing layer, which is usually an electron donor such as a conjugated polymer or a dye like a metal phthalocyanine and an electron acceptor. Up to now, fullerenes are the best acceptors for organic solar cells [Hoppe and Sariciftci, *J. Mater. Res.*, 2004]. As a matter of fact, they have an energetically deep LUMO, which induces a very high electron affinity relative to numerous organic donors. They exhibit relatively high electron mobility, up to 6 cm2/Vs [Anthopoulos et al., *Appl. Phys. Lett.*, 2006, Zhang, Domercq, and Kippelen, *Appl. Phys. Lett.*, 2007]. In the bilayer cells, charge separation predominantly occurs at the organic heterojunction. The built-in potential is determined by the HOMO/LUMO gap energy difference between the two materials contacting to form the heterojunction.

In order to improve the efficiency of planar herojunction structures, either p-i-n or stacked junctions have been probed. The p-i-n heterojunction structure consists of an active intrinsic region sandwiched between a n and a p-doped layer. [Niggemann et al., *Phys. Stat. Sol. (a)*, 2008]. As said above, usually a metal phthalocyanine is the p-type (electron donor) and fullerene the n-type (electron acceptor). The i photoactive layer can be a combination of both [Drechsel et al., *Organic electronics,* 2004].

As a conclusion of this chapter, in order to facilitate the discussion of the interface behaviour it is helpful to use an equivalent circuit model, which could be helpful in understanding of organic solar cells by providing a quantitative estimation for losses in the cells. The equivalent circuit commonly used to interpret the I-V characteristics of solar cells consists of a photogenerator connected in parallel with a diode, which represents the I-V characteristics in the dark. This corresponds to an ideal model in absence of parasitic resistances.

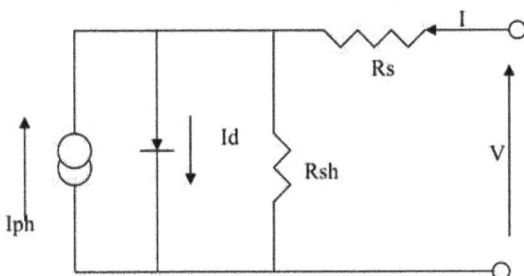

Figure 5 a

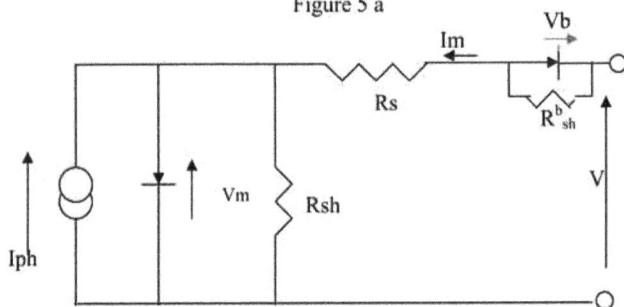

Figure 5 b

Fig. 5. Equivalent circuit of: a-solar cell with ohmic contacts, b-solar cell with rectifying back-contact.

However, in real organic solar cells, it is necessary to introduce a series resistance, Rs, and a shunt resistance, Rsh (Figure 5 a). For such solar cells the mathematical description of this circuit is given by the following equation:

$$I = I_0 \left[\exp\left(\frac{V - I \times Rs}{nkT} \right) - 1 \right] + \frac{V - I \times Rs}{Rsh} - Iph \qquad (2)$$

As discussed below, the values of Rs and Rsh depend on the properties of the electrode/organic interfaces.

4. Impact of electrode/organic interface properties on cells performances

As shown above organic optoelectronic devices are formed by sandwiching one or more semiconducting organic layers between conducting electrodes. The properties of the interface between the organic layers and the electrodes are determinant for the efficiency of the devices. In photovoltaic devices the contact must have good carrier extracting properties. Also electrodes must generate a built-in electric field that is sufficient to collect separated charges. Therefore a good understanding of the interface between the organic layers and the electrodes is crucial to achieve good organic optoelectronic devices.

When a π-conjugated organic material is put into contact with another material such as electrodes in optoelectronic devices the contact may result in different effects depending on the electrode properties. With a metal which surface is passivated the interaction results in a physisorption with possible integer electron charge transfer. If the metal is non reactive there is possibility of weak chimisorption and possible partial charge transfer. In the case of reactive metal, there is chemisorption with iono-covalent bonding between the metal and the organic material and corresponding partial charge transfert.

Often the surface work function of an electrode depends on the electrode history. The surface work function of a clean metal or TCO can be measured *in situ* by UPS for exemple (Φ_M = (5.2 eV for clean gold). However, after air, or only classical vacuum (10^{-3} Pa) exposure there is some physisorbtion at the metal surface which decreases its work function, for instance in place of the 5.2 eV expected, 4.5 eV are often obtained. Therefore air ambient exposure of gold results in a reduction by 0.7 eV of its work function [Braun, Salaneck, and Fahlman, *Adv. Mater.*, 2009]. Similar results have been obtained with TCO. For instance, Φ_{ITO} varies from 5 to 4.5 eV depending on the sample history [Li et al., *Thins Solid Films*, 2005; Johnev, et al., *Thins Solid Films* 2005]. So, the effective work function of the electrode when put in contact to an organic material may be significantly lower than that expected, which induces unexpected energy level alignment at the organic material/electrode interface. Often, in order to study the energy level alignment at the organic material/electrode interface UPS is used, and samples are grown *in-situ* to avoid contamination, that is to say all the process takes place under ultra-high vacuum. The result can fail in the attempt to describe the interface band alignment of optoelectronic devices contact, since they are processed under moderate vacuum, neutral gas or even room air conditions. That fact should be kept in mind when one attempts to understand the optoelctronic devices behaviour through UPS studies done in ultra high vacuum.

Energy level alignment at organic material/electrode interfaces is one of the main fundamental issues about the optoelectronic devices. In the light of inorganic semiconductor/metal interfaces the simple Schottky-Mott model has been often applied to organic contacts. The model is presented in Figure 6. When an organic semiconductor is put in contact with an electrode the Schottky-Mott model assumes an alignment of the vacuum level and a band bending in the space charge layer (SCL) to achieve alignment of the bulk Fermi levels. Therefore the barrier height for hole extraction Φ_B corresponds to:

$$\Phi_B = \Phi_M - \Phi_S \qquad (3)$$

Φ_S being the work function of the organic semiconductor (Figure 6).

More generally, if ϕ_M is the work function of the metal, i.e. its ionisation potential IP, and ϕ_S the work function of the semiconductor, i.e. the energy difference between the Fermi level and the vacuum level, for the semiconductor we will have:

- -in the case of n-type semiconductor
  - a Schottky-Mott contact if $\phi_M > \phi_S$, electrons diffuse from the semiconductor to the metal owing to small carrier density in semiconductors and to high carrier density

in metals, a depletion layer appears in the semiconductor: there is a rectifying contact.

- an ohmic contact if $\phi_M < \phi_S$, electrons diffuse from the metal to the semiconductor. There is a negative accumulation in the n-type semiconductor, no barrier forms, the contact is ohmic.
- -in the case of p-type semiconductor
  - a Schottky-Mott contact if $\phi_M < \phi_S$, holes diffuse from the semiconductor to the metal. Owing to small carrier density in semiconductors and to high carrier density in metals, a depletion layer appears in the semiconductor: there is a rectifying contact.
  - an ohmic contact if $\phi_M > \phi_S$, electrons diffuse from the semiconductor to the metal. There is a positive accumulation in the p-type semiconductor, no barrier forms, the contact is ohmic.

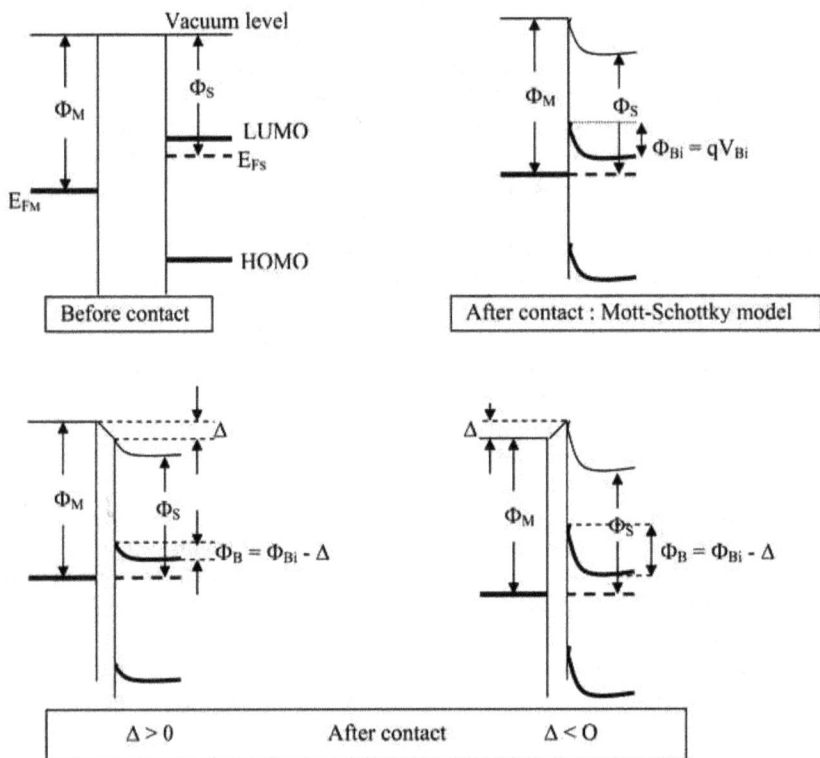

Fig. 6. Interface band alignment, before contact, after contact, without and with an interface dipole

Even if many studies have experimentally demonstrated a strong correlation between the metal work function and the barrier Φ_B for carriers exchange at electrode/organic material interfaces there is still matter of controversy.

The described Schottky-Mott model is an ideal and simple model, however, real energy level alignment should often consider a vacuum level discontinuity associated with an interface dipole, ID, resulting from charge rearrangement upon interface formation [Lee et al., *Appl. Phys. Lett.*, 2009].

In the case of inorganic metal/semiconductor contacts two limit models have been proposed. The Schottky-Mott model where the vacuum level of the organic and metal aligned, forming a region of net space charge at the interface and the Bardeen model, where a large density of surface states induces a pining effect of the Fermi level and the presence at the interface of a barrier independent of the metal work function. The Cowley-Sze model is an intermediate model, where interface states would be induced in the original band gap of the semiconductor upon contact with a metal giving the interfacial dipole Δ'. The effective barrier height for hole exchange $\Phi_{b,eff}$ is therefore given by :

$$\Phi_{B,eff} = \Phi_B - \Delta' \tag{4}$$

Δ' is proportional to the amount of charge transferred due to energy difference between the metal Fermi level and the charge neutrality level (CNL). If we assume a uniform distribution of metal-induced interface state, it can be shown that $\Phi_{B,eff}$ varies linearly with the metal work function with a slope, S, smaller than one [Lee et al., *Appl. Phys. Lett,.* 2009]. In the absence of metal-induced interface state, the injection barrier follows the Schottky-Mott limit with S = 1. The other limit corresponds to S = 0, the interface dipole reaches a saturated value with the organic CNL aligned to the metal's Fermi level. There is Fermi level pining and the variation of the metal work function is fully compensated by the metal-induced interface state dipole.

By analogy with inorganic metal/semiconductor contacts two limit models have been proposed when an organic semiconductor is deposited onto a conducting material. The first is the above described Schottky-Mott simple model. The second proposed that a charge dipole forms on the interface due to effect such as chemical interaction and/or formation of interface states, in that case the vacuum level does not align at the interface. This interface dipole (ID) induces vacuum level shift Δ. Therefore the Mott-Schottky barrier height should be modified by the amount of Δ:

$$\Phi_B = \Phi_M - \Phi_S - \Delta \tag{5}$$

The sign of Δ depends on the nature of the contact (Figure 6) and it will be discussed below. Moreover, another question is, does band bending occur in organic semiconductors? Following S. Braun and W.R. Salaneck, M. Fahlman [Braun, Salaneck, and Fahlman, *Adv. Mater.*, (2009)] band bending should not be expected for organic semiconductors, as they do not have band structure but localized state featuring hopping transport. Charge can be exchanged at the interface but only organic material in close vicinity to the metal surface takes part in the charge exchange. Yet, they admit that band-bending like behaviour has been demonstrated for π-conjugated organic thin films deposited on metal substrates. It has been shown that localized energy levels of the organic material are shifted depending on the distance to the metal interface, until depletion region thickness is reached [Nishi et al., Chem. Phys. Lett., (2005); Ishii et al., *Phys. Stat. Sol* (a), 2004]. Also, J. C. Blakesley and N. C. Greenham [Blakesley and Greenham, *J. Appl. Phys.*, 2009] have shown that there is a good agreement between UPS measurements and theoretical band bending calculations. UPS measurements of thin organic layers on conducting substrates have shown the presence of band bending within a few nanometers [Hwang et al., *J. Phys. Chem. C*, 2007]. It has been proposed that this band bending effect is due to transfert of carriers from the substrate into the organic film. Such integer charge transfer (ICT) at organic/passivated conducting substrate interface has been proposed by Salaneck group [Tengstedt et al., *Appl. Phys. Lett.* (2006); Fahlman et al., *J. Phys.: Condens. Matter*, (2009)]. The ICT model proposes that electron transfer via tunnelling through the passivating surface layer, which implies the transfer of an integer amount of charge, one electron at a time. Tunnelling occurs when the substrate work function is greater (smaller) than the formation energy of positively (negatively) charged states in the organic material. The energy of a positive integer charge transfer state E_{ICT+} is defined as the energy required to take away one electron from the organic material and, in the case of negative integer, the charge transfer state, E_{ICT-} is defined as the energy gained when one electron is added to the organic material. In the case of a positive integer charge transfer, the organic material at the interface becomes positively charged, while the substrate becomes negatively charged, creating an interface dipole Δ that down-shift the vacuum level. The electron transfer begins when the organic is put into contact with the substrate, and it goes on up until equilibrium is reached, i.e. when $E_{ICT+} \Delta$ is equal to the substrate work function (Figure 7).

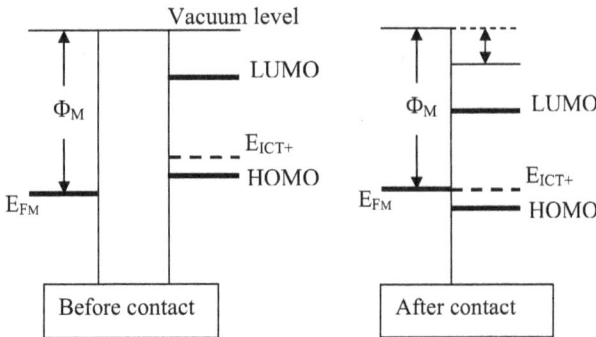

Fig. 7. Integer charge transfer model.

Here also there is some controversy about the formation, or not, of a band bending. However the model predicts the Fermi level pinning experimentally encountered when Φ_M < E_{ICT-} and Φ_M < E_{ICT+}, while it varies linearly with Φ_M between these two values [Tanaka et al., *Organic Electronics*, 2009].

In addition, Fermi level alignment is a critical problem. However in practical situation of organic solar cells, band bending coupled with interface dipole formation have demonstrated their potentiality to account for experimental results.

If the ICT model, with or without band bending, is efficient for passivated surface substrates other models should be used when there is some chemical interaction between the organic and the substrate.

In the case of strong chemisorption, for instance when the metal electrode is deposited by evaporation onto the organic material there is diffusion of metal atoms into the organic film and the situation is quite complicated, since often the organic material may offer different feasible bonding sites for the metal. Chemisorption can be used voluntarily to modify the properties of the substrate surface, typically by using self-assembled monolayers (SAM). SAM will be discussed in the paragraph dedicated to the contact anode/electron donor.

More generally, the chemical bonding between the metal and the organic molecule may involve a transfer of charge which up-shift, when there is an electronic charge transfer to the molecule, or down-shift, when there is an electronic charge transfer to the metal, the vacuum level by introducing a dipole-induced potential step at the interface (Figure 8). Therefore here also there is a shift Δ of the vacuum level at the interface.

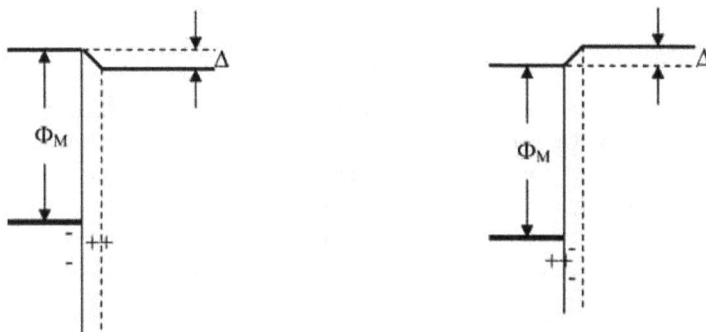

Fig. 8. Interface dipole involved by chemisorption's

As a conclusion it can be said that, whatever its origin, an interface dipole is often present at the interface electrode/organic. Following its sign, this dipole can increase or decrease the potential barrier present at the interface. However, this dipole is only one contribution to the interface barrier, the difference between the work function of the electrode (anode-cathode) and the energy level (HOMO-LUMO) of the organic material is another significant contribution, which allows predicting, at least roughly, the behaviour of the contact.

5. Interface characterisation techniques

One key issue for organic optoelectronic is the understanding of the energy-level alignment at organic material/electrode interfaces, which induces, a fortiori, the knowledge of the electrode work function and ionisation potential (HOMO) and electron affinity (LUMO) of organic semiconductors. For the investigation of the chemistry and electronic properties of interfaces X-ray photoelectron spectroscopy (XPS) and ultraviolet photoelectron spectroscopy (UPS) are often used [Braun, Salaneck, and Fahlman, *Adv. Mater.*, 2009]. Energy level alignment at organic/electrode interfaces can be also carefully studied with Kelvin probe [Ishii et al., *.Phys. Stat. Sol* (a) (2004)]. Cyclic voltammetry is also a valuable tool to estimate the HOMO and LUMO of the organic materials [Cervini et al., *Synthetic Metals*, 1997; Brovelli et al., *Poly. Bull.*, 2007].

5.1 Electron spectroscopy for chemical analysis (ESCA): X-ray photoelectron spectroscopy (XPS) and ultraviolet photoelectron spectroscopy (UPS)

ESCA is a widely used technique for studying chemical and electronic structure of organic materials. More precisely, the method is very useful for the study of surfaces and interfaces. In the case of UPS, the photoelectron inelastic mean free path is less than ten Angstroms. The well known basic equation used in interpreting photoelectron spectra is:

$$E_B = h\nu - E_{kin} - \Phi_{SP} \qquad (6)$$

Where E_B is the binding energy, $h\nu$ is the photon energy, ϕ_{SP} spectrometer specific constant (the work function of the spectrometer). Assuming that due to the removal of an electron from orbital i the rest of the electron system is not affected (frozen approximation), E_B corresponds to orbital energies $-\varepsilon(i)$. However, the remaining electrons in the environment can screen the photohole, which induces an additional relaxation contribution and impacts the measured E_B value. Changes in the valence electron density induces small, but significant, shift of the core level binding energy, called chemical shift. Hence, charge transfer and chemical bond formation can be probed using XPS. UPS is used for valence electronic study because the photoionisation cross-section for electrons is orders of magnitude higher in the valence band region for UPS and the photon source (He lamps) has high resolution. The source of photons is either HeI (hv = 21.2 eV) or HeII radiation (hv = 40.8 eV). These energies allow for mapping the valence electronic states of organic materials. The UPS spectra give information about the electronic structure of the material and its work function. It also measures the change Δ of the work function after coverage (Figure 9).

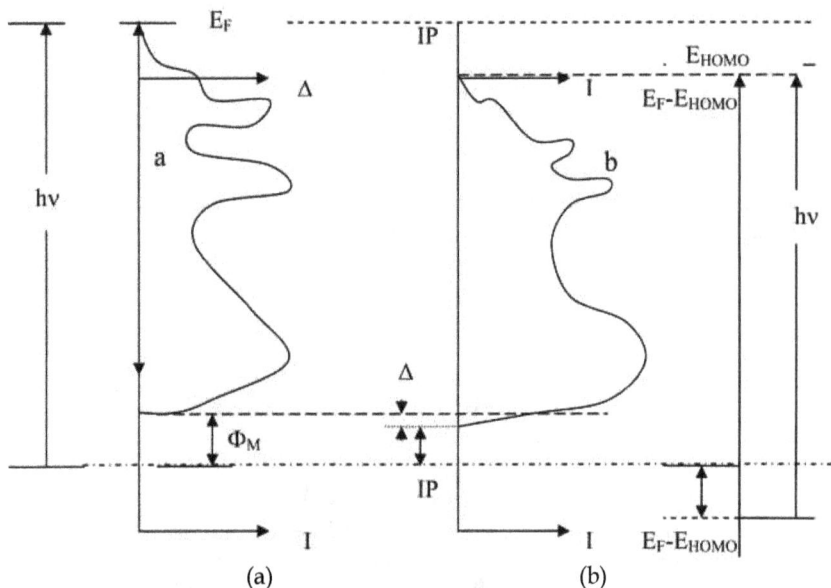

Fig. 9. Shows the principle of UPS for the study of an interface:
a- clean metal, b- metal covered with an organic monolayer.

The UPS spectrum of a clean metal substrate can be seen in Figure 9a. Electrons below the Fermi level are excited by the uv light and emitted into vacuum. The kinetic energy E_{kin} distribution of the emitted electrons is called the UPS spectrum and reflects the density of the occupied states of the solid.

Only photoelectrons whose kinetic energy is higher than the work function ϕ_M of a sample can escape from the surface, consequently ϕ_M can be determined by the difference between the photon energy and the width of the spectrum (Figure 9 a). The width of the spectrum is given by the energy separation of the high binding energy cutoff (E_{cutoff}) and the Fermi energy ($E_b = 0$):

$$\phi_M = hv - E_{cutoff} \tag{7}$$

A change in work function, Δ, then can be tracked by remeasuring the E_{cutoff} after deposition of an organic monolayer.

Possible shift of the cutoff and thus of the vacuum level suggests the formation of an interfacial dipole layer Δ [Crispin, *Solar Energy Materials & Solar Cells*, 2004; Kugler et al., *Chem. Phys. Lett.*, 1999; Seki, Ito and Ishii, *Synthetic Metals*, 1997] (Figure 9 b).

In this case the small binding energy onset corresponds to the emission from the highest occupied molecular orbital (HOMO) and the high binding energy (low kinetic energy) cutoff corresponds to the vacuum level at the surface of the organic layer.

Therefore as said above we can visualise the relative position of the energy levels at the interface, and examine the difference of the vacuum level between the metal and organic layer which corresponds to Δ (Figure 10).

UPS is a very powerful tool to detect the presence-or not- and to measure the interface dipole and therefore to understanding of the energy-level alignment at interfaces organic material/electrode.

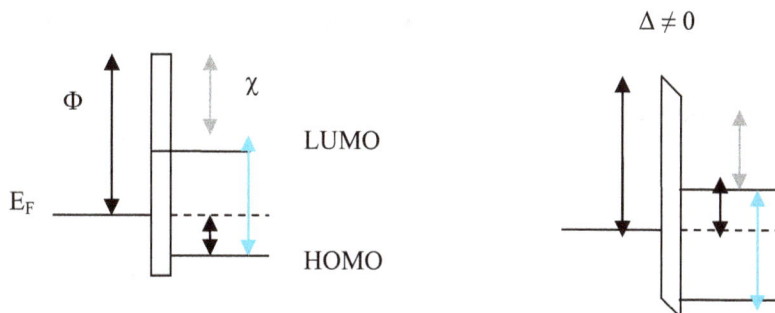

Fig. 10. Interfacial dipole Δ after contact: a: $\Delta = 0$, b: $\Delta \neq 0$.

4.2 Kelvin probe

The principle of Kelvin probe was put in evidence by Lord Kelvin in 1898 [Phil. Mag., 1898]. The principle was first applied, using a vibrating capacitor by Zisman [Zisman, *Rev. Sci. Instrum.*, 1932]. Nowadays, the Kelvin probe method (KPM) is used to measure the work function of various surfaces. The sample and a metallic vibrating reference electrode constitute a capacitor. The vibration of the reference electrode induces an alternative current, this current is zero when the voltage applied to the reference electrode is equal to the contact potential difference between the reference and the sample. When the sample is conductor, there is no difficulty, the surface of the sample works as a plate of the capacitor and charges are accumulated at the surface. It is more complicated when the sample is a semiconductor or an insulating material. Some part of the charge is into the sample, this situation has been discussed by different authors [Ishii et al., *Phys. Stat. Sol* (a), 2004; Pfeiffer, Leo and Karl, *J. Appl. Phys*, 1996]. They conclude that the vacuum level of the reference electrode exactly coincides with that of the sample, in the case of null-detection condition. Therefore it can be said that KPM probes the surface potential of the sample with precision.

For instance, the energy level alignment at CuPc/metal interfaces has been studied using KPM [Tanaka et al., *Organic Electronics*, 2009]. In order to study the vacuum level (VL) shift at CuPc/metal interfaces different metals presenting a wide range of Φ_M have been probed. Moreover, the deposition of the CuPc onto the metal was performed in a stepwise manner with Kelvin probe measurement at each step to follow the VL shift as a function of the CuPc film thickness. The study showed that the organic layer onto the metal surface plays two important roles in the energy level alignment: formation of an interfacial dipole (ID) and passivation of the metal surface. The deposition of the first nanometers (<2 nm) induces a large VL shift indicating a charge redistribution at the interface related to the interface dipole (ID) formation. For thicker thickness the VL variation depends on the Φ_M value. When Φ_M is higher than LUMO$_{CuPc}$ very little VL shift occurs for thicker films, the energy level alignment is determined by Δ_{ID} and Φ_M. Therefore the barrier height at the interface varies with Φ_M. When Φ_M is smaller than LUMO$_{CuPc}$, VL varies up to 5nm of CuPc, there is a spontaneous charge transfer (CT) from metal to the CuPc until LUMO$_{CuPc}$ is located above the Fermi level. There is a pinning of the Fermi level and the barrier height at the interface does not vary with Φ_M. This example shows the KPM could be an efficient tool for studying the interfaces organic materials/electrodes.

5.2 Cyclic voltammetry
Electrochemistry is a simple technique, which allows estimating the HOMO and LUMO of organic material [Li et al., *Synthetic Metals*, 1999].

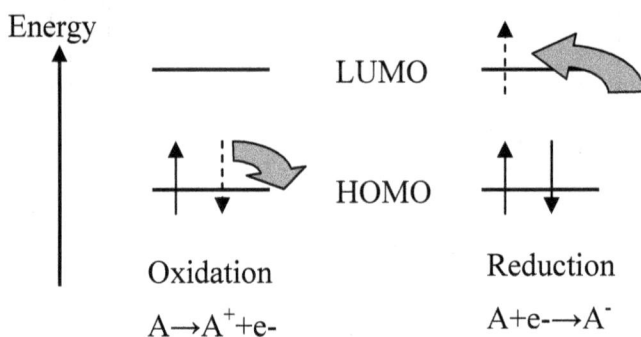

Fig. 11. Oxidation and reduction of an organic molecule.

When the organic material shows an electron reversible reduction and oxidation wave, cyclic voltammetry (CV) is recognised as an important technique for measuring band gaps, electron affinities (LUMO) and potential ionisations (HOMO). The oxidation process corresponds to removal of charge from the HOMO energy level whereas the reduction cycle corresponds to electron addition to the LUMO (Figure 11).

The experimental method is based on cyclic voltammetry [Cervini et al., *Synthetic Metals*, 1997; Li et al., *Synthetic Metals*, 1999.]. The electrochemical set up was based on classical three electrodes cells. The reference electrode was Ag/AgCl.

The electrochemical reduction and oxidation potentials of the organic material are measured by cyclic voltammetry (CV). When the CV curves showed a one electron reversible reduction and oxidation wave, the HOMO and LUMO energy can be determined from the first oxidation and reduction potential respectively. The potential difference Eg = LUMO – HOMO can be used to estimate the energy gap of the dye. The energy level of the normal hydrogen electrode (NHE) is situated 4.5 eV below the zero vacuum energy level [Brovelli et al., *Poly. Bull.,* 200)]. From this energy level of the normal hydrogen and the reduction potential of the reference electrode used, for example Ag/AgCl i.e. 0.197 V versus NHE, a simple relation can be written which allows estimating the both energy values (7):

$$LUMO = [(-4.5)-(0.197-Ered)]eV.$$

$$HOMO = [(-4.5)-(0.197-Eox)] eV. \qquad (8)$$

As an example the curve corresponding to N,N'-diheptyl-3,4,9,10-perylenetetracarboxylicdiimide (PTCDI-C7) is presented Figure 12.

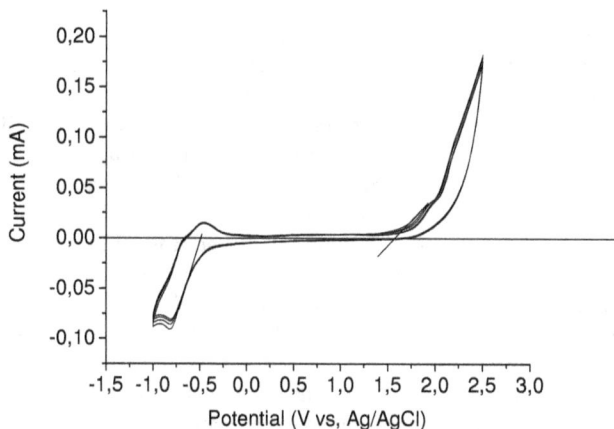

Fig. 12. Cyclic voltammogram of PTCDI-C7 on Pt disc electrode in medium of anhydride dichloromethane.

As working electrode, a polycrystalline platinum disc was used. The reference electrode was Ag/AgCl in solution of tetraethylammonium chloride (Et_4NCl). The potential was adjusted to 0.199 mV with respect to the normal hydrogen electrode (NHE) [East and del Valle, *J. Chem. Educ.,* 2000]. As counter-electrode, spiral platinum was used in a separated compartment of work electrode by fritted glass

The electrochemical reduction and oxidation potentials of the PTCDI-C7 were measured by cyclic voltammetry (CV) (see Figure 12). From CV curves, PTCDI-C7 in dichloromethane showed a one electron reversible reduction and oxidation waves.

The HOMO and LUMO energy of PTCDI-C7 can be determined from the first oxidation and reduction potential respectively. The potential difference Eg = LUMO-HOMO can be used to estimate the energy gap of the dye. The energy level of the normal hydrogen electrode (NHE) is situated 4.5 eV below the zero vacuum energy level [Bard and Faulkner, *Fundamentals and Applications, Wiley* 1984]. From this energy level of the normal hydrogen and the reduction potential of the reference electrode used in the present work Ag/AgCl i.e. 0.199 V versus NHE, a simple relation allows us to estimate the both energy values:

$$LUMO = [(-4.5)-(0.199-Ered)] eV$$

$$HOMO = [(-4.5)- (0.199-Eox)] eV \qquad (9).$$

The values of oxidation and reduction potential are 1.57 V and –0.38 V respectively. Relatively to the vacuum level the energy values of HOMO and LUMO levels are –6.30 eV and –4.30 eV respectively. Therefore the band gap estimated from the electrochemical measurements is 2.0 eV. This value is only slightly higher than the optical band gap of a PTCDI-C7 thin film (1.95 eV). So, the energy gap calculated from the difference between the LUMO and HUMO energies is quite close to the optical band gap, which testifies that the cyclic voltammetry provides a useful rough estimate for the location of the LUMO and the HOMO of the organic materials.

6. Interface organic acceptor/cathode

For electron injection (OLED) or collection (solar cells) it is necessary to incorporate a low work function as cathode. However low work function metals such as Mg, Li, Ca... are not suitable because they have high reactivity in air. Historically works on OLEDs have shown that aluminium coupled with LiF is a very efficient cathode. Hung et al. [Hung, Tang and Mason, *Appl. Phys. Lett.* 2008] have shown that when an ultra thin (1 nm) LiF layer is deposited onto the organic material before Al, this LiF/Al bilayer cathode greatly improved the electron injection and reduced the threshold voltage.

The increase in luminance and efficiency is attributed to enhancement of the electron injection from the aluminium into the organic acceptor. The LiF/Al cathode improves injection by raising the Fermi energy and shifting the effective injection interface deeper into the organic film [Baldo and Forrest, *Phys. Rev.*, 2001.]. Effectively there is Li doping of the organic layer during Al deposition.

In the case of solar cells, insertion of a thin LiF layer (< 1.5 nm) at the organic/aluminium interface allows improving the power conversion efficiency of the cells. An increase in the forward current and in the fill factor is observed upon reducing the serial resistivity across the contact. The optimum LiF thin film thickness is around 1 nm. For higher values the high resistivity of the LiF decreases its beneficial influence. From (I-V) curves it has been estimated that the insertion of a thin LiF layer decreases the serial resistivity of the diodes by a factor 3-4, while the shunt resistivity is stable [Brabec et al., *Appl. Phys. Lett.*, (2002).]. The precise mechanism of LiF on the interface properties is still under discussion. Moreover, it should be highlighted that, in the case of solar cells, LiF is not as successful as in the case of OLEDs. Therefore a lot of works have been dedicated at the improvement of the organic acceptor/cathode interface. Different buffer layers have been probed and the main results are summarized below.

We have seen that the maximum value of Voc is Voc $\leq$ LUMO$_A$ – HOMO$_D$. The same dependence of Voc with LUMO$_A$ – HOMO$_D$ has been encountered whatever the structure used, bulk heterojunction or multiheterojunction structures. The same controversy on the dependence of Voc with the cathode work function [Chan et al., *Appl. Phys. Lett.*, 2007; Rand, Burk and Forrest, *Phys. Rev.*, 2007] is present for both structure families. Indeed, if the Voc value is effectively related to Δ(LUMO$_A$ – HOMO$_D$), it depends also of others parameters such as the dark current (leakage current), Voc decreases when this current increases, that is to say when the shunt resistance, Rsh, is faint (Figure 5). In order to check the variation of Voc with Δ(LUMO$_A$ – HOMO$_D$) and Rsh, we have studied a cell family with the structure ITO/Donor/Acceptor/Al/P, with donor = ZnPc or CuPc, acceptor = C$_{60}$, PTCDA, PTCDI-C7 and 1,4-DAAQ and P a protective layer from oxygen and humidity contamination, which allows keeping the device in room air after assembling. P$_I$ corresponds to an encapsulation before breaking the vacuum and P_A an encapsulation after 5 min of room air exposure [Karst and Bernède, *Phys. Stat. Sol. (a)*, 2006]. While in the former case there is no aluminium post depot oxidation, at least during the first hours of air exposure, in the latter case, 5 min of air exposure induces air diffusion at the grain boundaries of the polycrystalline Al layer and formation of a thin Al_2O_3 between the anode and the organic material.

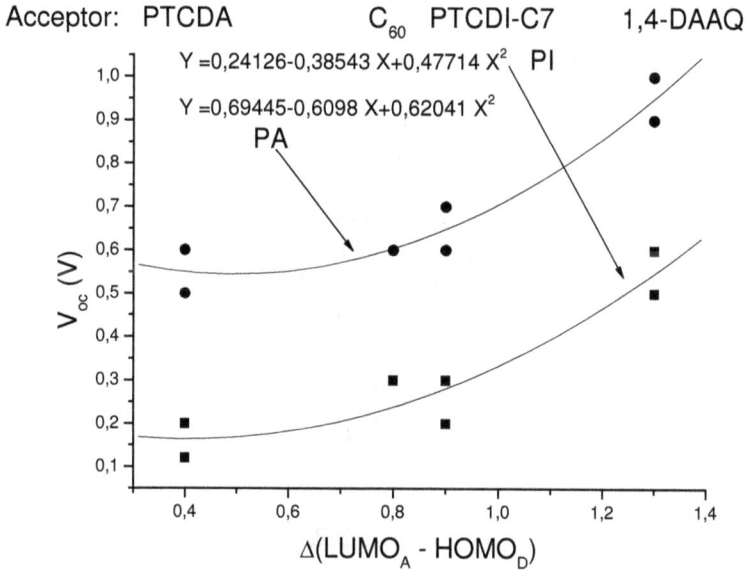

Acceptor: PTCDA C_{60} PTCDI-C7   1,4-DAAQ

$Y = 0,24126 - 0,38543\ X + 0,47714\ X^2$ PI

$Y = 0,69445 - 0,6098\ X + 0,62041\ X^2$

PA

V_{oc} (V)

$\Delta(LUMO_A - HOMO_D)$

Fig. 13. Voc variation with $\Delta(LUMO_A - HOMO_D)$.

The results are summarized in Figure 13. It can be seen that, as expected, the Voc value increases with the $\Delta(LUMO_A - HOMO_D)$. However, it can be seen also that two curve families are clearly visible. One with small Voc values, which corresponds to cell encapsulated without breaking the vacuum and another with higher Voc values, which corresponds to cells encapsulated after 5 min of air exposure. The two curves are nearly parallel, which demonstrates that the same phenomenon is at the origin of the Voc increase. Since the only difference between these two families is the contact or not with room air, the translation of the curve should be attributed to the presence of the thin natural Al_2O_3 layer at the electron acceptor/aluminium interface. This natural oxide does not depend on the organic material but only on the alumium electrode air exposure, which is in good agreement with the translation effect of the two curves. Such ultra thin Al_2O_3 layer (1nm) increases the shunt resistance value, which justifies the Voc value increase. Such effect of aluminium oxidation on the open circuit voltage has already been proposed by Singh and coll.[Singh et al., *Appl. Phys. Lett.*, 2005; Singh et al., *Sol. Energy Mater. Sol. Cells*, 2006], thanks to our in situ encapsulation process we have directly put this effect in evidence. However, if the increase of the shunt resistance of the cells through insulating oxide formation at the interface cathode/organic materiel, allows increasing the open circuit voltage and therefore the solar cells efficiency, the limit of the positive effect of such oxide layer is rapidly achieved. Indeed, it is only efficient when electrons can tunnel through the oxide layer. Beyond 2.5 nm, not only the shunt resistance increases but also the series resistance and therefore the current and cell efficiency.

Moreover other limitation at the interface organic/cathode has been highlighted through the experiments described below. It has been shown that one way for circumventing the diffusion length limitation is to use cells with multiple interfaces. Peumans et al. [Peumans et al., *Appl. Phys. Lett.*, 2000] have shown that the introduction of a thin large band gap organic material allows improving significantly the device performances. He called electron blocking layer (EBL) this thin film, because its bandgap was substantially larger than those of the organic donor and acceptor, which block excitons in the organic semiconducting layer far from the cathode avoiding any quenching effect at the cathode/organic interface. Will see more precisely the effect of this "EBL", but first we will conclude on the effectiveness of the very thin oxide layer between the cathode and the organic electron acceptor. In order to discriminate between the effect of an EBL and an oxide layer deposited before the cathode we have worked with ITO/CuPc/C_{60}/Alq3/Al/P, Alq3 being used as EBL layer. It is shown in Table 1, that, as expected, the EBL improve significantly the cells performances, while the encapsulation process does not modify the strongly the I-V characteristics.

Devices	J_{SC} (mA/cm^2)	Voc (V)	Rsh (Ω)
ITO/CuPc/C_{60}/Al/PI	4.75	0.24	90
ITO/CuPc/C_{60}/Al/PA	4.40	0.41	1650
ITO/CuPc/C_{60}/Alq3/Al/PI	7.75	0.45	1800
ITO/CuPc/C_{60}/Alq3/Al/PA	7.45	0.48	1850

Table 1. Jsc and Voc values of the different devices under AM1.5 conditions.

In fact, the Voc value in the presence of Alq3 does not depend strongly on the encapsulation process, while it does when simple CuPc/C_{60} junction is used. This difference can be explained by the variation of the value of the shunt resistance, Rsh. Without Alq3, a thin Al_2O_3 layer is necessary to improve Rsh and therefore Voc, with Alq3, Rsh is sufficient and the alumina is not necessary to optimise the Voc value (Table 1).

Accordingly to the present discussion, the EBL is sufficient to confine the photogenerated excitons to the domain near the interface where the dissociation takes place and prevents parasitic exciton quenching at the photosensitive organic/electrode interface. Also it limits the volume over which excitons may diffuse. For vapor deposited multilayer structures, a significant increase in efficiency occurs upon the insertion of the exciton blocking interfacial layer, interfacial layer, between the cathode and the electron acceptor film. Bathocuproine (BCP) is often used as exciton blocking buffer layer [Peumans et al., *Appl. Phys. Lett.*, 2000; Huang et al., *J.Appl. Phys.*, 2009]. However, with time, BCP tends to crystallize, which induces some OSCs performance degradation [Song et al., *Chem. Phys. Lett.*, 2005]. Consequently, either other more conductive [Refs] or more stable, e.g, aluminium tris(8-hydroxyquinoline) (Alq3), materials have been tested as EBL [Song et al., *Chem. Phys. Lett.*, 2005; Hong, Huang and Zeng, *Chem. Phys. Lett,.*, 2006; Bernède and al., *Appl. Phys. Lett.*, 2008]. Therefore, many organic materials with quite different HOMO and LUMO values can be used as EBL. Indeed, it appears that EBL can also protect the electron accepting film from atoms diffusion during deposition of the electrode. Also it is thick enough and sufficiently homogeneous to fill pinholes and others shorting effect which increases Rsh and therefore Voc and the cell efficiency. Therefore the EBL protects the fragile organic films from damage produced during electrode deposition onto the organic material. The large band gap of the EBL, larger than that of the adjacent organic film, allows blocking the excitons in this film. If the EBL blocks the excitons it should not block all charge carriers. Therefore the EBL should be chosen so that it allows electrons collection at the cathode. However the offset energy of the highest occupied molecular orbital (HOMO) of the electron donor (often the fullerene) and the EBL (such as the bathocuproine) is large. Moreover, the optimum EBL thickness is around 8 nm, which is too thick to allow high tunnelling current. So, even if the EBL is an electron conducting material, the difference of the LUMO levels of C_{60} and Alq_3 implies that electrons must overcome a large energy barrier to reach the Al cathode in case of electron transport via LUMO levels (Figure 14-1). More probably, the charge transport in the EBL is due to damage induced during deposition of the cathode, which introduces conducting levels below its LUMO (Figure 14-2) and explains the reason why the transport of electron is not weakened. [Rand et al., *Adv. Mater.*, 2005]. As a conclusion, the EBL, not only block the excitons far from the cathode where they can be quenched, but also prevents damage of the electron acceptor film during cathode deposition. It should be transparent to the solar spectrum to act as a spacer between the photoactive region and the metallic cathode and it must transport electrons to avoid high series resistance. The EBL is also important for fabricating large-area devices with a low density of electrical shorts.

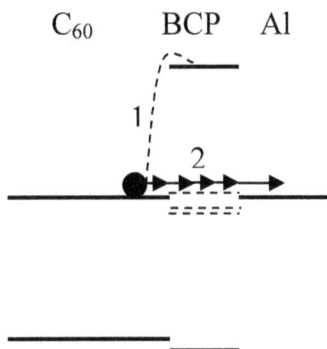

C_{60} BCP   Al

Fig. 14. Band schemes of organic films and cathode contact.

7. Anode/organic donor interface

Globally, the electrodes in contact with the organic materials have great importance in the device behaviour. Of course, in optoelectronic devices it is necessary to allow the maximum amount of photons of the solar spectrum to enter the active part of the device. Therefore it is necessary that at least one of the electrodes should exhibit high transparency and should not be reflecting. In classical configuration the transparent electrode, a transparent conductive oxide thin film (TCO), is the anode. Typically, glass coated with the degenerate semiconductor indium-tin oxide (ITO), is used as anode electrode. ITO, which commonly serves as anode in organic optoelectronic devices, attracted considerable interest due to its unique characteristics of high conductivity, good transparency in the visible domain and easy patterning ability, moreover it is commercially available.

A crucial point in organic devices is the interface between the inorganic electrodes and the organic materials. The key parameter at the anode interface is the hole collection from the organic semiconductor to the anode. A barrier for carrier transport is often present at the interface. It is usually determined by the electrode/organic band offset, that is to say, in the case of holes, the difference between the work function of the anode and the highest occupied molecular orbital (HOMO) of the electron donor, even if, as discussed above, the barrier height depends also of the presence, or not, of an interface dipole. About the influence of the barrier height at the interface anode/organic donor Kang et al [Kang, Tan and Silva, *Organic Electronics,* 2009] have shown a clear relation between the work function of the anode and the devices performances. They show that the energy conversion efficiency of the cells follows the variation of the value of the anode work function. The work function was measured by Kelvin probe, the anode, ITO/PTFE (polytetrafluoroethylene), was treated with different UV exposure time. The work function increases during the first five minutes and then decreases, also the devices performances and mainly the short circuit current Jsc. High Jsc in organic solar cells are mainly due to small barrier height between the anode and the organic and subsequently improved carriers extraction process. Therefore, the influence of the barrier height at the contact anode/electron donor being well established, it is necessary to control the work function of the anode to achieve good band alignment and ohmic contact. High work function anode is desirable to decrease the series resistance. High and reproducible work functions are difficult to obtain for ITO [Bruner et al., *J. Am. Chem. Soc.,* 2002]. Many processes have been proposed to achieve this goal. First, as discussed in paragraph 3, it should be underlined that ITO work function depends strongly of the thin film history and it is quite difficult to predict. It has been shown that ITO surface chemistry is difficult to control, because its surface is covered by hydrolysed oxides [Armstrong et al., *Thin Solid Films,* 2004), Donley et al., *Langmuir* (2002); Kim, Friend & Cacialli, *J. Appl. Phys.,* 1999]. In fact, he surface chemical functionality of ITO is not well understood [Katkova et al., *Appl. Surf. Sciences,* 2008]. Authors propose the presence of hydroxyl [Purvis et al., *J. Am. Chem. Soc,.* (2000)] others not [Chaney & Pehrsson, *Appl. Surf. Sci.,* 2003]. What is clear for all experimenters in the field of organic optoelectronic devices is that cleanliness of the ITO surface is critically important for efficient hole exchange at the organic material/ITO interface. Devices performances, not only depend on the surface treatment but also on the deposition batches [Berredjem et al., *The European Physical Journal: Applied Physics,* 2008].

Moreover, it is well known that crystals in polycrystalline ITO thin films have pyramidal shape, which induces a significant surface roughness (some nm) of these films. This surface roughness is often evocated as a source of leakage current and lifetime limitation in optoelectronic devices. Also, ITO electrodes were reported to interact chemically, which contributes to the degradation of optoelectronic devices performances [Kugleret al., *Synthetic Metals,* 1997)]. For instance, even in the absence of oxygen and moisture, oxidation of organic material in contact with ITO has been reported [Scott et al., *J. Appl. Phys.*1997]. It appears that ITO anode serves as source of oxygen. At least, it should be underlined that ITO, is not ideal due to the scarcity of its main component: indium. Indeed, to day, ITO is widely used as electrode in optoelectronic devices and demand for indium is expected to outstrip supply these years, making devices based on ITO expensive. All that justifies, not only the different works dedicated to ITO surface treatment itself, but also original works on different TCOs and transparent anodes.

First of all different surface treatments of ITO have been probed. Hydrogen peroxide treatment improves the devices performance through work function increase ($4.7 < \Phi_M < 4.8$

eV), however, even with similar Φ_M, different turn-on voltage are measured. Obviously, additional factors should be considered such as surface roughness performance [Kugler et al., *Synthetic Metals*, 1997]. Different acidic solutions have been probed (HCl, H_3PO_4).

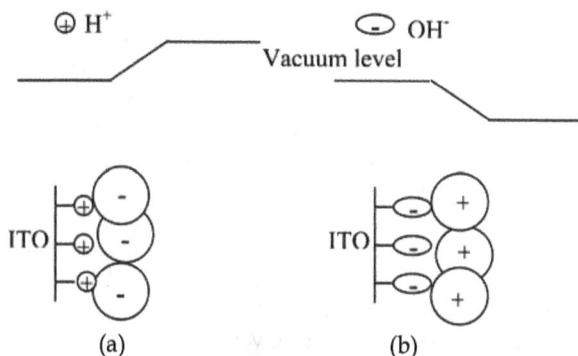

Fig. 15. Surface structure of passivated ITO for acid (a) and base (b) treatment.

Treatments with phosphoric acid lead to an increase in work function of about 0.7 eV (4.5 to 5.2 eV) with good homogeneity [Johnev et al., *Thin Solid Films*, 2005].

Such effect is induced by monolayer adsorption (Figure 15 a), which allows improving the solar cells efficiency from 1.2 to 1.5 %.

However this efficiency remains smaller than that obtained with conducting polymer buffer layer, which will be discussed below. When treated by a base a decrease of Φ_M is obtained, which means that Φ_M can be shifted of 1 eV [Nüesch et al., *Appl. Phys. Lett,.* 74 1999]. Moreover, it is necessary to use an appropriate plasma treatment before chemical adsorption. As a matter of fact, plasma treatments are often used to increase the ITO work function. In the case of plasma treatment, after chemical pre-cleaning, the sample was treated by RF plasma, usually the ambient gas is Ar or O_2, with better results achieved with O_2. Not only the plasma treatment cleans the ITO surface, increases Φ_M, but also smoothes the film surface, whole things resulting in performance improvement of devices [Lu & Yokoyama, *Journ. Crys. Growth*, 2004, Zhong & Jiang, *Phys. Stat. Sol.* (a), 2006]. Another well known technique used to tune the ITO surface work function, is the deposition of self-assembled monolayers (SAM) onto the ITO film surface. A SAM consists of a molecular backbone terminated by an anchoring group and, at the other extremity, by an end group that may induce a dipole, Δ_{SAM}. Δ_{SAM} is defined as positive if it up-shifts the vacuum level on the organic/SAM side. Different organic material families can be used as SAM, tin phenoxides [Bruner et al., *J. Am. Chem. Soc.*, 2002], thiophene phosphonates , phosphonic acids [Hansson et al., *J. Am. Chem. Soc.*, 2005; Sharma et al., *J. Appl. Phys.*, 2009], also polymeric (LBL) assembly has been used for anode modification layer by layer [Kato, *J. Am. Chem. Soc.*, (2005].

If these chemical techniques allow tuning efficiently the work function of the ITO thin films, physical techniques such as spin coating, vapor deposition can be used also with success.

The conducting polymer the most widely used to help the charge transporting at the interface ITO/organic is the poly(ethylene dioxythiophene) doped with polystyrene sulfonic acid (PEDOT:PSS) [Hoppe and Sariciftci, *J. Mater. Res.*, 2004]. PEDOT:PSS is a p-type semiconductor, a good hole transport material, it is soluble in water and easy to depose by spin coating. Its work function is 5.2 eV.

The initial solution of PEDOT-PSS is 3 wt. % in water. It is spun at 2000-5000 rpm to form a 50-100 nm thick layer. After deposition, to prevent the presence of water in the device, PEDOT:PSS coated ITO is annealed for half to an hour at 100-150°C. Then the different organic constituents and the cathode of the optoelectronic device are deposited. The PEDOT:PSS buffer layer allows the device performance to be significantly improved, OLEDs and solar cells. It is admitted that the high value of its work function allows a good band alignment with the HOMO of the electron donor, which decreases the barrier height at the interface and therefore assures a better hole collection from the polymer into the ITO electrode. Also it is supposed that the PEDOT:PSS spin coated onto the ITO surface smoothes its surface and, therefore, any possible short circuiting due to the spiky roughness of the ITO surface is prevented. It improves the contact between the polymer and the ITO. It is admitted that this buffer layer enhances adhesion to the organic layer. Also it prevents direct contact between the oxygen of the ITO and the organic material.

However, PEDOT:PSS is problematic since its poor conductivity is a major limiting factor for device performance and it degrades under UV illumination [Chang & Chen, *Appl. Phys. Lett,.* 2007; Kang et al., *J. Phys. D: Appl. Phys.*, 2008]. Even after baking, due to its hygroscopic nature some amount of water always appears in PEDOT:PSS, which introduces water into the active layer, it is also slightly acidic [Van de Lagemaat et al., *Appl. Phys. Lett.*, 2006, Johnev et al., *Thin Solid Films,* 2005]. Moreover, not only the depositing process from aqueous solution introduces impurities but the reproducibility is in need of improvement [Johnev et al., *Thin Solid Films,* 2005].

Therefore, other solutions have been proposed, each one based on original buffer layers such as metal or oxides. Some attempts using thin metal buffer layers have been done during the last years, however the results were quite disappointing, the metal thin film used being thick of some nanometers, the transmission of the visible light decreases significantly (Figure 16) and also the devices efficiency [Yoo et al., *Synthetic Metals,* 2005].

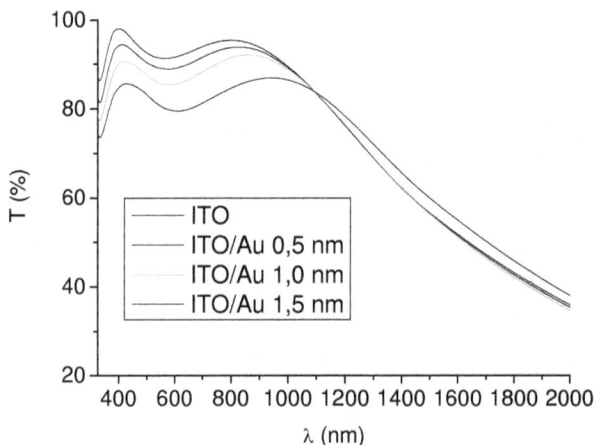

Fig. 16. Variation of the transmittance of ITO/Au structures with the Au thickness (0 to 1.5 nm).

We have shown that this difficulty could be overcome by using an ultra thin (0.5 nm) gold film. The introduction of this ultra-thin metal layer at the interface anode/electron donor allows improving significantly the energy conversion efficiency of the organic solar cells [Bernède et al., *Appl. Phys. Lett.*, 2008; Bernède et al., *Sol. Energy Mater. Sol. Cells*, 2008].

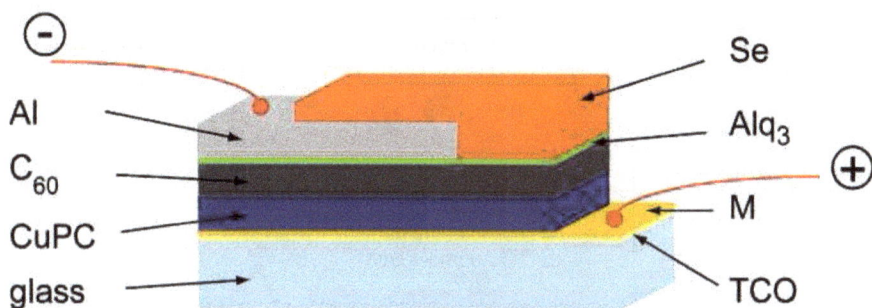

Fig. 17. Schematic structure of the fabricated solar cells with the ultra-thin gold layer onto the TCO.

The efficiency improvement is even more remarkable as regards to TCO initial quality. The effect of this ultra-thin metal buffer layer has been probed on multi-heterojunction organic solar cells (Figure 17) and we present, with more details, this example of efficient buffer layer at the interface anode/electron donor. The electron donor used was copper phthalocyanine (CuPc) (some attempts have been done using pentacene and similar behaviour has been obtained), the electron acceptor was fullerene (C_{60}) and the electron blocking layer was the tris(8-hydroxyquinoline) (Alq_3) [Kim et al., *Sciences*, 2007, Berredjem et al., *Eur. Phys. Journ.: App. Phys.* 2007]. CuPc, C_{60} and Alq_3 have been deposited in a vacuum of 10^{-4} Pa. The thin film deposition rate and thickness were estimated in situ with a quartz monitor. The deposition rate and final thickness were 0.05 nm/s and 35 nm in the case of CuPc, 0.05 nm/s and.40 nm in the case of C_{60} and 0.1 and 9 nm for Alq_3. These thicknesses have been chosen after optimisation.

After organic thin film deposition, the aluminium upper electrodes were thermally evaporated, without breaking the vacuum, through a mask with 2 mm x 8 mm active area. This Al film behaves as the cathode, while the ITO is the anode. Some ITO anodes have been covered with an ultra thin metal film deposited by vacuum evaporation, the metal being Au, Cu, Ni. The thickness of these ultra thin metal films, M, has been varied from 0.3 to 1.2 nm. Finally, the structures used were: glass/ITO(100nm)/M ($0 \leq x \leq 1.2$ nm)/CuPc(35nm)/C_{60}(40nm)/Alq_3(9 nm)/Al(120nm). It can be seen in figure 18 and table 2 that the presence of the ultra-thin gold layer improves significantly the solar cells performances. When different batches of ITO were used, without Au buffer layer, the solar cells performance vary strongly, while they were of the same order of magnitude when an ultra-thin gold layer was deposited onto ITO (Table 2).

Similar results have been obtained when AZO and FTO are used. The performances of organic solar cells using this ultra thin metal layer, are nearly similar, whatever the TCO used [Bernède et al., *Appl. Phys. Lett.*, 2008, Bernède et al., *Sol. Energy Mater. Sol. Cells*, 2008]. This suggests that indium free organic devices with high-efficiency can be achieved, which can contribute to the sustainable development.

Batch	Anode	Jsc (mA/cm²)	Vco (V)	FF	η %
a	ITO	7.31	0.45	0.44	1.45
	ITO/Au (0.5nm)	8.09	0.49	0.56	2.25
b	ITO	6.80	0.40	0.25	0.67
	ITO/Au (0.5nm)	8.34	0.45	0.50	1.86

Table 2. Photovoltaic performance data of devices achieved with batches a and b of ITO, under AM1.5 conditions.

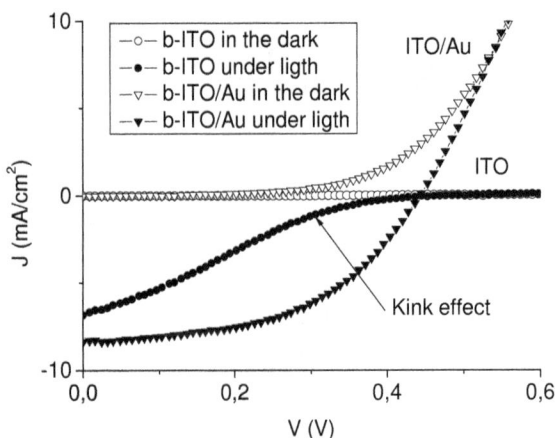

Fig. 18. Typical J-V characteristics of solar cells, with an anode of ITO (batch b) covered (▼) or not (●) with 0.5 nm of Au, in the dark (open symbol) and under illumination of AM1.5 solar simulation (100 mW/cm²) (full symbol).

As said above, others metal such as Cu, Ni, Ag, Pt have been probed, however, up to now, the best results have been obtained with a gold ultra thin film. It should be noted that, roughly, the organic solar cells performance increases with the metal work function, which means that, Ag which have the smallest Φ_M gives the worst performances.

In order to understand the ultra-thin gold layer effect, TCO covered with such gold layers have been characterized by scanning electron microscopy (SEM), atomic force microscope (AFP), X-ray photo-electron spectroscopy, optical transmission measurement (Figures 16, 19). It is shown that the ultra-thin film is discontinuous, while the roughness of the TCO/Au electrode is not different from that of bare TCO (0.8 nm). However it can be seen through the XPS study that the CuPc has grown more homogeneously when deposited onto gold covered TCO. Moreover as shown in Figure 18 the shape of the J-V characteristic depends on the anode configuration, in the case of bare TCO a "kink" effect is clearly visible, while classical diode characteristics are obtained when the TCO is covered by the ultra-thin gold layer.

In order to discuss the effect of the ultra thin gold film on solar cells performances, we recall shortly supposed beneficial effect of the classical buffer layer, the PEDOT:PSS.

As said above, up to now, the most common buffer layer inserted at this interface is the PEDOT:PSS, its contribution to the improvement of solar cells performance has been attributed to:

a-ITO (rms = 0.80 nm)

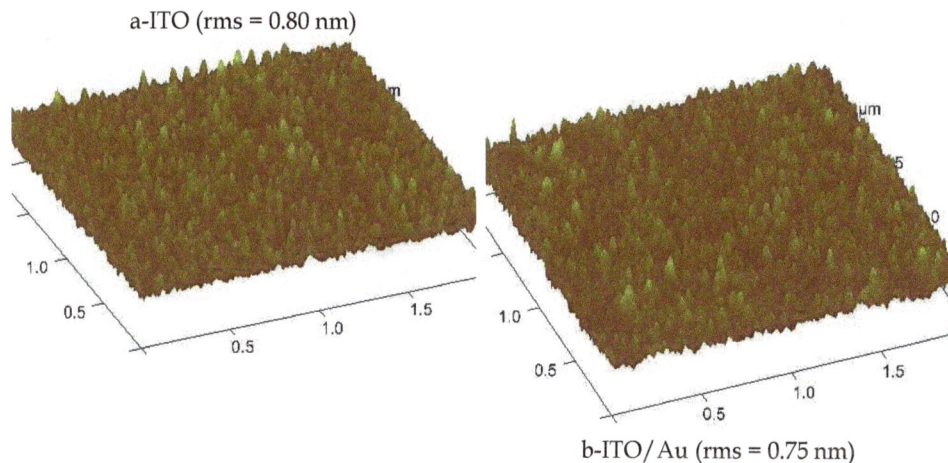

b-ITO/Au (rms = 0.75 nm)

Fig. 19. AFM images of Au (0.5 nm) covered ITO (a) and bare ITO (b).

- the smoothing effect of the quite rough TCO surface, therefore, any possible short circuiting due to the spiky roughness of the TCO surface is prevented.
- the physical separation, which avoids direct contact between the oxygen of the TCO and the organic material.
- its work function ($\phi_{M(PEDOT:\ PSS)}$ = 5.1 eV), which allows to decrease the barrier height at the interface, since the work function of the TCO is smaller (4.5-4.7 eV for ITO), while the HOMO of CuPc is 5.2 eV.
- the improvement of the contact between the polymer and the TCO. It is admitted that this buffer layer enhances adhesion to the organic layer.

It can be seen immediately that the two first contributions can be excluded when an ultra-thin gold film is substituted by PEDOT:PSS, since the roughness of the modified anode is similar to that of the initial TCO and the gold film is discontinuous. The two last contributions seem more probable. The work function of gold is 5.1 eV and therefore the ultra-thin gold layer can improve the matching between the work function of the anode and the highest occupied molecular orbital (HOMO) of the organic electron donor (Figure 5a, b, c). Such contribution will be discussed more carefully below. The fourth contribution is in good agreement with the XPS study, which shows that the CuPc films are more homogeneous when deposited onto gold modified TCO.

Therefore from the examples presented above it can be concluded that the two main contributions to the interface improvement by inserting an ultra-thin metal buffer layer between the TCO anode and the CuPc are a better matching of the band structure (Figure 20) and a higher homogeneity of the organic film.

The equivalent circuit model (figure 5 a) could be helpful in understanding of organic solar cells by providing a quantitative estimation for losses in the cells. As said above, the equivalent circuit commonly used to interpret the I-V characteristics of real solar cells consists of a photogenerator connected in parallel with a diode and a shunt resistance, and a series resistance. For such solar cells the mathematical description of this circuit is given by the equation (1).

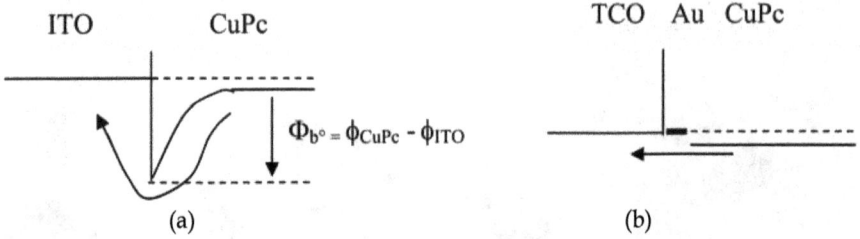

Fig. 20. Contact anode (ITO)/ electron donor (CuPc)
a-without buffer layer; b-with a gold buffer layer.

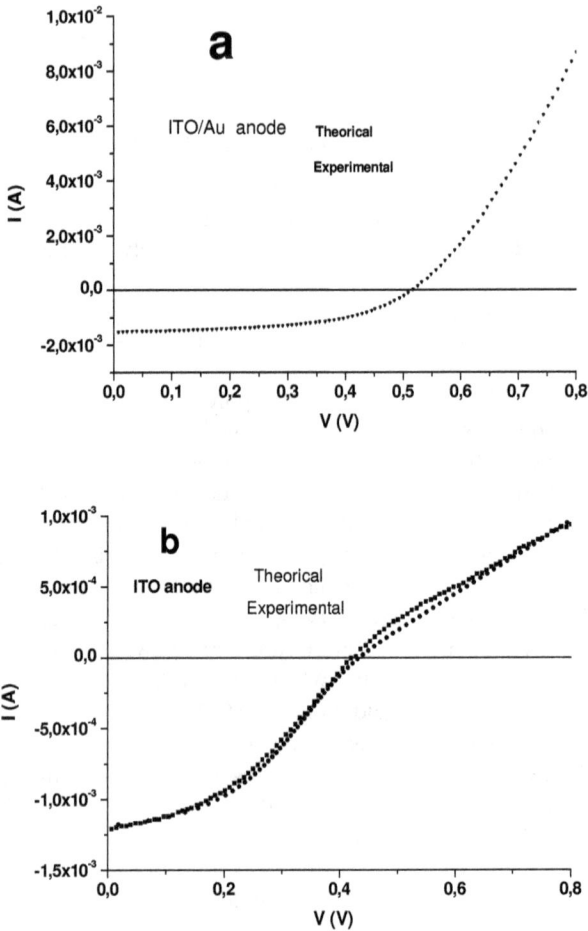

Fig. 21. I-V characteristics under AM1.5 illumination of a solar cell using a ITO/ Au anode
(a) and a ITO anode (b), (•) experimental and (■) theoretical curves.

In a recent contribution [Kouskoussa et al., *Phys. Stat. Sol.* (a), 2009], we have shown that the Lambert W-function method can be used to determinate Rs, the series resistance, Rsh the shunt resistance, n the ideality factor of the diode and Iph the photo-generated current.

The problem to be solved is the evaluation of a set of five parameters Rs, Rsh, n, Iph and Is in order to fit a given experimental I-Vcharacteristics using a simple diode circuit.

A good agreement between the experimental and theoretical fitted curves is achieved with ITO/Au anode (Figure 21 a), while it is not in the case of bare ITO anode.

As said above no agreement could be achieved in the case of a bare ITO or Ag covered anode, whatever the series and shunt resistance proposed. Such impossibility shows that the simple equivalent scheme used in this theoretical study cannot explain the experimental results obtained with a bare ITO anode.

In the case of bare and Ag covered ITO anode, it is necessary to assume the presence of a back-contact barrier at the ITO/CuPc interface (figure 5 b), to achieve a good fit between experimental and theoretical results. Assuming a thermoionic current at this interface, the hole current is:

$$I_b = - I_{b0} (\exp(-qV_b/kT)-1) \tag{10}$$

With: - I_{b0} saturation current,
  - V_b voltage across the back contact,
  - k Boltzmann constant,
  - T temperature.

Therefore the current-limiting effect, "rollover", is due to the back-contact barrier height. It occurs because the total current saturate at a value J_{b0} [Demtsu and Sites, *Thin Solid Films*, 2006]. The value of J_{b0} is the current value where the J-V curve starts to show rollover.

Demtsu and Sites have treated the main junction and the back-contact junction (Figure 6 b) as independent circuit element. Here, when a forward bias V is applied to the circuit, the voltage is divided between V_m across the main CuPc/C_{60} junction, V_b across the back-junction TCO/CuPc and IRs across the series resistance:

$$V = V_m + V_b + IRs.$$

Under illumination the current across the main junction is:

$$I_m = I_{m0} (\exp(qV_m/nkT)-1) - Iph + Vm/Rsh \tag{11}$$

And through the back contact it is:

$$I_b = - I_{b0} (\exp(-qV_b/kT)-1) + Vb/ R^b_{sh} \tag{12}$$

Equating equations (10) and (11):

$$I_{m0} (\exp(qV_m/nkT)-1) - Iph + Vm/Rsh + I_{b0} (\exp(-qV_b/kT)-1) - Vb/ R^b_{sh} = 0 \tag{13}.$$

The parameters Rs and I_{m0}, n, Rsh of the main diode are calculated in the region far from the saturation current Ib. As said above, I_{b0} is the current value where the J-V curve starts to show rollover. Then equation (13) can be solved.

A good agreement can be achieved between experimental and theoretical curves, (Figure 21 b), which validates the hypothesis of the presence of a rectifying effect at the ITO/CuPc contact.

The parameters extracted are reported in Table 3. It should be noted that when ITO is covered with Ag, two diodes are necessary to obtain a good fit between the experimental and theoretical curves. Since the work function of Ag is only 4.3 eV, this result is in good agreement with the present discussion. One can see in Table 3 that, when the two diodes model is used, the estimated Rs values are of the same order of magnitude as those of ITO/Au anode. Therefore the introduction of a back junction diode is a good interpretation of the interface ITO/CuPc. It justifies the small fill factor and short circuit current values when bare ITO is used as anode in the heterojunction solar cells. The Voc value can be improved thought an increase of the shunt resistance value. It can be seen in Table 3 that the shunt resistance value of the cells with a ITO/Au anode is significantly higher.

Anode		Main junction CuPc/C_{60}			Back-contact junction	
	n	Rsh	Rs	I_{mo} (A)	$R^b{}_{sh}(\Omega)$)	I_{bo}(A)
ITO	2.55	330	20	2 10^{-6}	15000	2.58 10^{-6}
ITO/Ag	2.6	400	40	1 10^{-6}	13200	2.24 10^{-6}
ITO/Au	1.5	6900	25	2.23 10^{-9}	-	-

Table 3. Parameters calculated using a main diode and (or not) a back contact diode.

Table 3 shows that the ideality factor, n, decreases, while Rsh increases, when the ITO anode is covered by an ultra-thin Au layer, which corresponds to a significant improvement of cell performance. When Ag is used the improvement is not significant. The increase of Rsh can be related to the growth of pinhole free organic films when deposited onto Au coated TCO [Kim et al., Sol. Energy Mater. Sol. Cells, 2009].

When thicker metal films are introduced at the anode/organic interface, surface plasmon effect has been proposed as alternative approach toward enhanced light absorption without the need for thick films. [Chen, et al., Appl. Phys. Lett., 2009]. A surface plasmon is an optically generated wave, which propagates along a metal/dielectric interface. In tuning the light excitation, a resonance can occur when the frequency of the incident photon equals the collective oscillation frequency of conduction electrons of metallic particles. These properties can be used in the photovoltaic domain in order to improve the light absorption. Theoretical calculations have been performed to determine optimal plasmonic materials to optimise light absorption [Duche, et al. Sol. Energy Mater. Sol. Cells, 2009]. Enhanced absorbance up to 50% has been experimentally obtained in a 50 nm thick blend film including silver nanospheres with a diameter of 40 nm. Devices based on BHJ structures using ITO anode covered with Ag nanodots permits 20% improvement of the efficiency [Kim et al., Appl. Phys. Lett., 2008]. The plasmon resonance wavelength depends on the metal nanodots properties, it can be tuned by annealing the silver film or using other metal [Morfa,. Appl. Phys. Lett., 2008]. Even if the efficiency of the surface plasmon effect in the field of solar cells is always under discussion, it is often proposed to improve the light absorption.

After metals, different oxides (MoO_3, WO_3, ZnO) have been probed as buffer layer at the TCO anode /CuPc interface.

In the case of OLEDs [Im et al., Thin Solid Film, 2007; Hsu & Wu, Appl. Phys. Lett., 2004; Shi, Ma, and Peng, Eur. Phys. J. Appl. Phys., 2007; Qiu et al., J. Appl. Phys., 2003; Matsushima et al., Appl. Phys. Lett., 2007; Jiang et al., J. Phys. D: Appl. Phys., 2007; You et al., J. Appl. Phys., 2007] and organic thin film transistors [Chu et al., Appl. Phys. Lett., 2005; Park, Noh and Lee, Appl. Phys. Lett., 2006], different attempts have been done, with some success, using different oxide thin films as buffer layer between the TCO and the organic material. Also some attempts have been probed in the case of organic solar cells [Yoo et al., *Synthetic Metals*, 2005; Chan et al., *J. Appl. Phys.*, 2006]. The use of such buffer layer is based on the idea that the potential barrier at the ITO/organic donor interface can be reduced by elevation of ITO surface work function, giving high hole transport at this interface. In order to check this hypothesis, using the same multiheterojunctions as above, we have probed different buffer layers at the ITO/CuPc interface. The buffer layers used were MoO_3, WO_3 and ZnO. The resulting different I-V characteristics are compared to those of a reference cell built on untreated anode. The different energy band diagrams are proposed and discussed. Finally, we demonstrate that the reduction process of the barrier at the ITO/organic donor interface depends on the type of material: oxide or metal.

During this study, the majority of the ITO anodes have been covered with an oxide buffer layer. These thin buffer layers were deposited by vacuum evaporation. The MoO_3 oxide thin films were thermally evaporated onto the ITO coated glass. The thickness of these ultra thin oxide films ranged between 1 and 7.

Finally, the structures used were: glass/ITO(100nm)/bufferlayer/CuPc(40nm)/C_{60}(40nm)/Alq_3(9nm)/Al(120nm)/P_{Se}.

The optimum efficiencies were achieved for thickness of around 3.5 ± 1 nm, that means that, as in the case of gold, the oxide film does not completely cover the anode surface since it has been shown that 5 nm are necessary to obtain continuous thin MoO_3 film, using thermal evaporation [Song et al., *Chem. Phys. Lett.*,2005]. For thicker oxide films, if the fill factor (FF) stays far higher than the value obtained with bare ITO, the short circuit current (Jsc) is slightly smaller and it progressively decreases as the oxide thickness increases. With thinner oxide films the inflection point, in the J-V characteristics, typical of the bare ITO anode (Figure 22), is still slightly present.

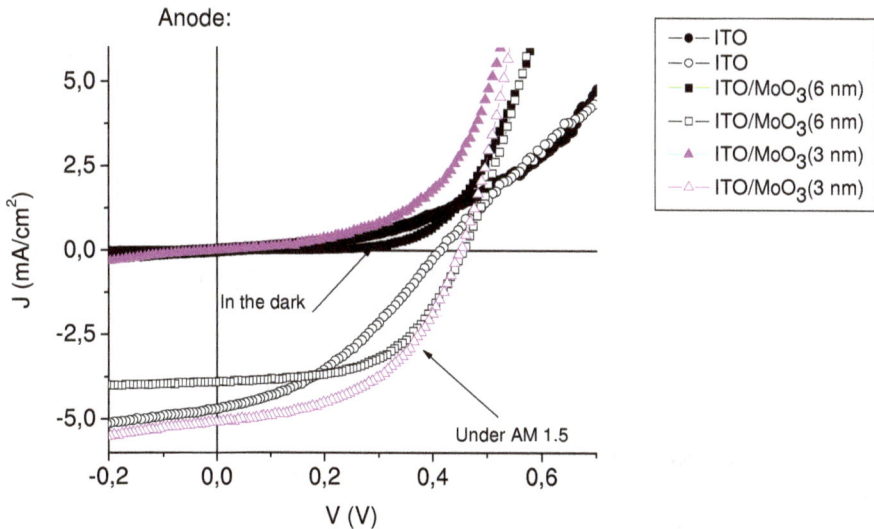

Fig. 22. Typical J-V characteristics of Anode/CuPc/C_{60}/Alq_3/Al structure, with Anode = ITO (●), ITO/MoO_3 (6 nm) (■) and ITO/MoO_3 (3 nm) (▲), in the dark (full symbol) and under illumination of AM1.5 solar simulation (100 mW/cm²) (open symbol).

It should be highlighted that the main improvement induced by the oxide is related to the fill factor (Table 4). Here also the improvement induced by the oxide does not depend on the TCO used, since the best result achieved by the cells deposited during the run corresponds to FTO. Here also the presence, or not, of the kink effect testifies of the efficiency of oxide.

A similar approach to the one used in the case of gold buffer layer can be used in the case of oxide buffer layer. Often the positive effect on the properties of the organic optoelectronic devices of the anode buffer layer is mainly attributed to the reduction of the barrier energy between the ITO, which is usually the anode, and the organic layer (electron donor for solar cells, hole transporting layer for OLEDs...).

Anode	ITO	ITO/MoO_3 (6 nm)	ITO/MoO_3 (3 nm)
Jsc (mA/cm²)	4.69	3.90	5.05
Voc (V)	0.41	0.46	0.45
η (%)	0.73	0.97	1.13
FF (%)	38	54	49.7

Table 4. Photovoltaic performance data under AM1.5 conditions of devices using ITO, ITO/MoO_3 (3 nm) and ITO/MoO_3 (6 nm) anodes.

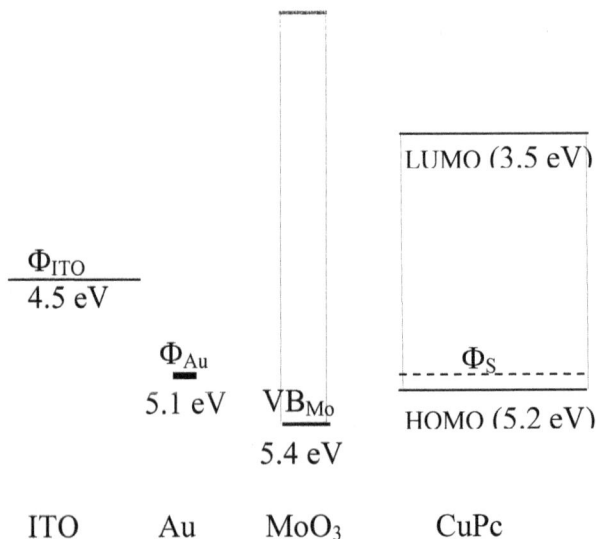

Fig. 23. Band scheme before contact

If the buffer layer material exhibits a high work function value (Figure 23), i.e., a work function in good agreement with the HOMO of the organic layer, it is often suggested that there is a simple alignment of the energetic levels of the anode and the organic material, without any discussion of the electrical properties of the buffer layer. However the electrical behavior of conductive, semi conducting and insulating materials is very different and therefore the nature of their interfaces will depend on the their properties.

If we introduce an insulating film, (I), between the metal, (M) (or degenerated semiconductor), and the semiconductor, (S), the contact behaves like a MIS structure. This means that there is not band alignment between the TCO, the insulating layer and the organic semiconductor but, qualitatively, the band diagram of Figure 24. This results from the fact that the Fermi level must be constant throughout the metal/insulating layer/semiconductor and therefore [Demtsu et Sites, *Thin Solid Films*, 2006]:

$$\Phi_{b^\circ} = \Phi_b + \Delta \qquad (15)$$

The distribution between Φ_b and Δ depends on the insulating properties of the oxide and the semiconductor properties. The space charge, which forms in the depletion layer of the semiconductor, can be expressed as an equivalent surface state density Qsc. In the ideal case, that is to say in the absence of any space charge effect in the interfacial insulating layer, an exactly equal and opposite charge Qm develops on the metal surface. In this case of an ideal insulating layer the potential Δ across the interfacial layer can be obtained, following the application of Gauss law to the surface charge on the metal and semiconductor:

$$\Delta = \delta(Qm/\varepsilon_i) \qquad (16)$$

ε_i being the dielectric constant of the insulating layer and δ its thickness [Sze, Physics of Semiconductor Devices, John Wiley Editor, 1981]. Therefore, even if there is not a simple alignment of the work function of the insulating film with those of the conductive anode and of the semiconductor, the presence of this layer can significantly modifies the barrier value at the interface through the potential Δ [Park, Noh and Lee, *Appl. Phys. Lett*, 2006]. As a summary, in the case of the present work, since we have (Figure 23): $\Phi_{ITO} = 4.5$ eV, $\Phi_{Au} = 5.1$ eV, $HOMO_{CuPc} = 5.2$ eV, $VB_{MoO3} = 5.4$ eV.

- the ITO/CuPc contact induces a pseudo Schottky contact, $V_{bi} = \phi_M - \phi_S$ between 0.5 and 0.7 eV, CuPc being an electron donor (p-type semiconductor), ϕ_S is slightly smaller than $HOMO_{CuPc}$ (Figure 20 a), while the Au/CuPc, and therefore ITO/Au/CuPc, contacts induce a nearly ohmic contact (Figure 20 b).
- the ITO/MoO$_3$/CuPc behaves like a MIS structure, that is to say the insulating layer induces a decrease of the barrier height with $\Phi_b = \Phi_{b^\circ} - \Delta$ [Cowley and Sze, *J. Appl. Phys.*, 1965]. As discussed above the efficiency of the insulating layer in the band alignment depends on the insulating properties and the thickness of the MoO$_3$ layer (Figure 24).

We have now all the keys necessary to explain the behaviour of the different cells studied in the present work. It is well known that a barrier is present at the interface ITO/organic material (electron donor in the case of solar cells and hole transporting layer in the case of OLEDs). The introduction of an interfacial layer between the ITO and the organic material allows to decrease the barrier height at the interface, which facilitate the hole collection (solar cells) or hole injection (OLEDs). It is clear from the discussion above that an oxide or a metal with sufficiently high work function value should be used.

MoO$_3$ has already been used to increase the hole injection in OLEDs [Matsushima et al., *Appl. Phys. Lett.*, 2007; Jiang et al., *J. Phys. D: Appl. Phys.*, 2007]. The decrease in the barrier height, allowed by the introduction of a MoO$_3$ layer, increases with its thickness (equation 16). However, as shown by the experimental study, there is an optimum thickness value.

When the film is too thin ($\leq$ 1.5 nm) the Δ value is small and moreover the covering efficiency of the ITO by the MoO_3 is not complete. For MoO_3 layers thicker than 6 nm, the short circuit current significantly decreases (Figure 22).

In Figure 24 it is supposed that the carriers cross the insulating layer by tunnel effect. The interfacial buffer layer is assumed to be a few angstroms thick and transparent to carriers whose energy is greater than the potential barrier of the semiconductor. However, when the thickness of the insulating layer increases, the probability of tunnel effect decreases. It is admitted that up to 3 nm the efficiency of the tunnel effect is maximum, while it decreases progressively for thicker films.

We have shown that the optimum MoO_3 thickness is around 3.5± 1 nm, which means that the insulating layer is efficient even for thickness higher than the theoretical optimum value. However evaporated molybdenum oxide films are strongly oxygen deficient and Rozzi et al have shown, in a theoretical study, that, when some oxygen is removed from the MoO_3 crystal, some Mo4d antibonding orbitals located in the gap are filled up [Rozzi et al., *Phys. Rev.*, 2003]. Therefore holes can cross the insulating film by multiple tunnelling steps [A.G. Milnes, D.L. Feucht, "Heterojunctions and metal semi-conductor junctions.", Academic Press Editor, 1972.] effect through these gap states introduced by the oxygen vacancies. As a result, the probability of tunnelling though the barrier increases. Finally the optimum MoO_3 thin film thickness, 3.5± 1 nm, corresponds to a compromise between an optimum ITO coverage and a sufficient transparency to charge carriers.

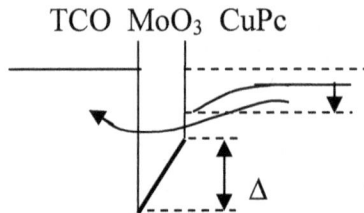

Fig. 24. Band scheme after contact: Hole transfer from the organic electron donor to the anode for different anode configurations after contact.

8. Example of the effect of an organic buffer layer used in different configurations

Factors which limit organic solar cells performance include limited spectral sensitivity, carriers separation at the interface acceptor/donor, low carrier mobility, energy step at the organic material/electrode. In devices based on ED/EA heterojunction, the theoretically obtainable open circuit potential (Voc) is given by the difference between the highest occupied molecular orbital of the electron donor ($HOMO_D$) and the lowest unoccupied molecular orbital of the electron acceptor ($LUMO_A$): $\Delta(HOMO_D - LUMO_A)$ [Zimmermann et al., *Thin Solid Films* 2005]. The introduction, at the electron donor /electron acceptor interface, of an ultra-thin organic layer could allows controlling the Voc value by modifying this difference as shown by Kinoshita and col. [Kinoshita et al., Appl. Phys. Lett., 2007]. They use with success this multicharge separation interface concept to increase the Voc value. They introduce an ultra-thin CuPc layer, which is usually used as electron donor, at the pentacene/fullerene interface. CuPc has also been proved to be an efficient buffer layer [Hong, Huang, and Zeng, *Chem. Phys. Lett.,* 2006]. In order to improve carrier mobility, structural templating of CuPc has been obtained using an ultra-thin 3,4,9,10-perylenetetracarboxyl dianhydride interlayer [Sullivan, Jones and Ferguson, *Appl. Phys. Lett.,* 2007]. In the present example, similar experiments have been done with a perylene derivative:N,N'–diheptyl-3,4,9,10-perylenebiscarboximide (PTCDI-C7).

8.1 Synthesis of the N,N'–diheptyl-3,4,9,10-perylenebiscarboximide (PTCDI-C7)

PTCDI-C7 (Figure 25) was synthesized by condensation of PTCDA with heptylamine [Demmig and Langhals, *Chem. Ber.,* 1988]. This compound is sufficiently soluble in chloroform to allow a chromatographic method for purification. Yield: 85 %. The elemental microanalyses results are in good agreement with the expected ones (weight %: C = 77.76; H = 6.74; N = 5.11).

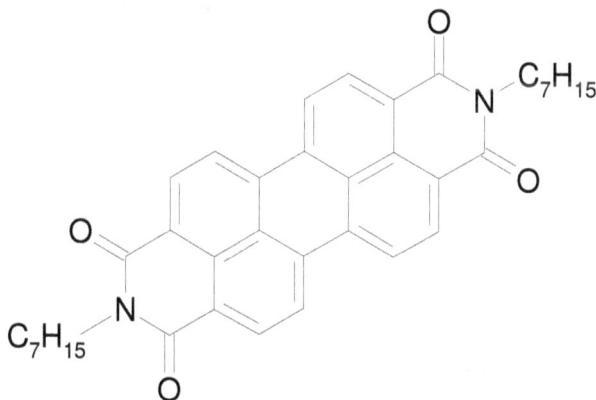

Fig. 25. PTCDI-C7 molecule.

8.2 Organic solar cells realisation

In the present series of experiments, the TCO electrode was a layer of indium tin oxide (ITO) on glass substrate. CuPc, PTCDI-C7, C_{60} and Alq_3 have been deposited in a vacuum of 10^{-4} Pa. The thin film deposition rate and thickness were estimated *in situ* with a quartz monitor. The deposition rate and final thickness were 0.05 nm/s, 35 nm for CuPc, 0.05 nm/s and 40 nm in the case of PTCDI-C7 and or C_{60} and 0.05 nm/s and 8 nm for Alq_3. The thicknesses have been chosen after optimisation.

We have shown earlier that the presence of an ultra-thin metallic film at the TCO/organic material interface allows improving significantly the power conversion efficiency of organic solar cells [Bernède et al., *Appl. Phys. Lett.,* 2008]. So, a very thin Au film (0.5 nm) has been introduced at the interface ITO/CuPc in the most cases. The thickness of the organic films has been checked with a scanning electron microscope (SEM). After organic thin film deposition, aluminium upper electrodes were thermally evaporated (150 nm thick), without breaking the vacuum, through a mask with 1.5 mm x 6 mm active area.

8.3 Experimental tests and discussion of the effect of PTCDI-C7 buffer layers

As justified above, PTCDI-C7 has been probed as a buffer layer at the CuPc/C_{60} and anode/electron donor interfaces. Following the multicharge separation interfaces concept, in order to improve Voc, an ultra-thin PTCDI-C7 has been introduced at the heterojunction $CuPc/C_{60}$ interface. The PTCDI-C7 thickness was 4 nm. However, while there was a strong decrease of the short circuit current Jsc, there was not any Voc increase (Figure 26). It has already been shown [Zhou et al., *Sol. Energy Mater. Sol. Cells*, 2007] that, in the case of bulkheterojunctions, charge separation yield achieved with perylene derivatives is smaller than the one achieved with C_{60}. The same effect can explain the strong Jsc decrease in the presence of PTCDI-C7 at the $CuPc/C_{60}$ interface.

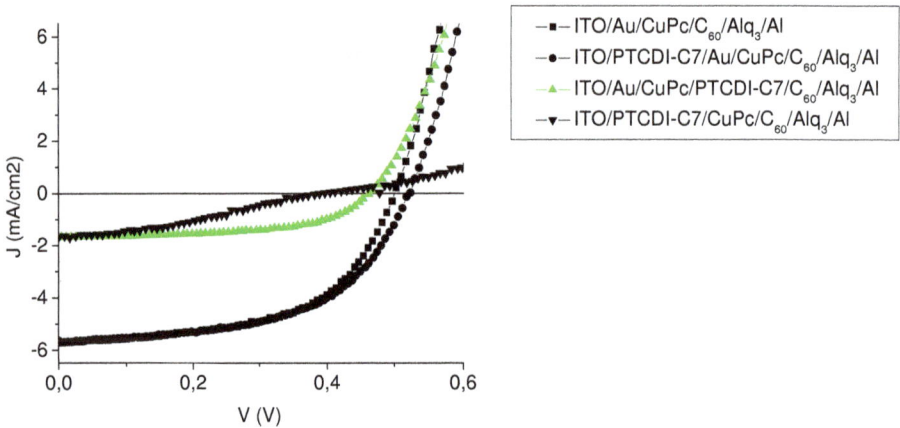

Fig. 26. Typical J-V characteristics under illumination of AM1.5 solar simulation (100 mW/cm²) with different anodes.

Then PTCDI-C7 has been probed as buffer layer at the anode/electron donor interface. A thin PTCDI-C7 ultra-thin layer has been introduced at the anode/CuPc interface. The results are summarized in Figure 26 and table 5. It can be seen that, if, as expected, the presence of Au improves significantly the devices performances, it is not the case when PTCDI-C7 is substituted to Au. However, when a PTCDI-C7 ultra-thin film (0.2nm) is deposited onto ITO and then covered with Au (0.5 nm) there is a significant increase of Voc. Such increase is reproducible. A systematic study of the thickness of this buffer layer has shown that 0.2 nm is the optimum value, for thicker PTCDI-C7 film there is a fast decrease of Jsc, FF and therefore of the efficiency of the cells (Table 5).

Au thickness (nm)	0.5	0.5	0.5	0.5	0.5	0	0
PTCDI-C7 thickness (nm)	0	1	2	3	4	2	0
Voc (V)	0.50	0.50	0.52	0.48	0.47	0.1	0.44
Jsc (mA/cm²)	5.63	5.62	5.64	3.75	2.88	0.35	3.21
FF %	57	56	56	34	26	22	0.24
η %	1.60	1.58	1.63	0.61	0.35	0.008	0.33

Table 5. Photovoltaic characteristics, under light AM1.5 (100 mWcm⁻²), for different cells using PTCDI-C7 as buffer layer deposited onto the ITO

Sullivan and col. [Sullivan, Jones and Ferguson, *Appl. Phys. Lett.*, 2007] have shown that, in the case of PTCDA buffer layer, there is some structural templating of CuPc which results in a higher carrier mobility and therefore Jsc increase. It can be observed in Figure 27, that the CuPc film grown on PTCDI-C7 exhibits a slightly higher ordering than those grown onto ITO or ITO/Au. The tendency of CuPc to grow with higher degree of order is expected to improve the carrier mobility in the films.

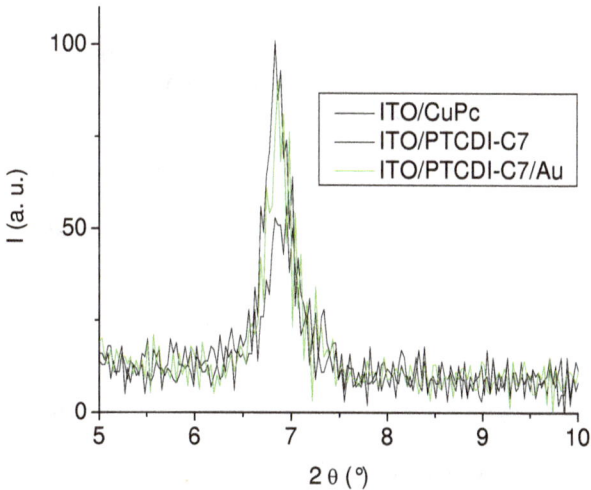

Fig. 27. Difractograms of CuPc thin films, 35 nm thick, deposited onto different cathodes

However, in our study, the more significant effect of the PTCDI-C7 buffer layer is not an increase of Jsc but of Voc. Probably, since the presence of the PTCDI-C7 buffer layer induces higher degree of order of CuPC films it should avoid the presence of pinholes and other defects growth, which increases the shunt resistance value of the solar cells. We have already shown that such diode improvement allows increasing the Voc value [Karst and Bernède, *Phys. Stat Sol. (a)*, 2006]. The study of the PTCDI-C7 films has shown that it has small conductivity. When the film thickness is only 2 nm, the carriers can cross the buffer layer by tunnel effect and the photocurrent is not decreased, for thicker buffer layer, the tunnel effect efficiency decreases and does the current.

In conclusion, the example of PTCDI-C7 developed above is a good illustration of the very broad possibility of original buffer layers. Here, a first gold buffer layer allows, as shown above to improve significantly the solar cells performances mainly through a decrease of the barrier height at the interface organic donor/TCO anode. Besides, the gold buffer layer increases the shunt resistance by decreasing the density of pinholes present in the CuPc films. We have shown here that this last effect could be improved by inserting a thin PTCDI-C7 at the interface just onto the TCO. This layer induces structural templating of CuPc as shown by the XRD study. Therefore, while PTCDI-C7 is clearly not an ideal surface modification of the anode in term of energy level alignment for hole collection (cf 4.3), the effect of shunt resistance increase allows improving the open circuit voltage.

9. Conclusion, toward the future.

In the last 22 years that have elapsed since the pioneering work of Tang [Tang, *Appl. Phys. Lett.*, 1986], significant improvement in the fundamental understanding and cells construction have led to efficiencies higher than 6%. The new concept of polymer:fullerene BHJ solar cells has allowed dramatic improvements in devices efficiency. It has induced a healthy competition with the multi-heterojunction devices base on small organic molecules, which induces significant progress in both cells families. A common aspect of both families is the great difficulty to control the electrode/organic material interface. This aspect is largely developed in the present work. One key issue for organic optoelectronic is the understanding of the energy-level alignment at organic material/electrode interfaces and the knowledge of the influence of the bottom electrode on the structural properties of the organic films. Even if not easy to predict, since dipole effect related to different possible origin are often present, the behaviour of an electrode/organic material contact can be roughly estimated using quite classical theory such as the Schottky-Mott model. Therefore the knowledge of the HOMO and LUMO values of the organic materials and that of the work functions of the electrodes are very helpful in the choice of new materials and structures configuration.

The challenges for the next organic solar cells improvements focus not only on new efficient absorbing organic materials, but developing new buffer layers, new cells architecture, new transparent conductive electrodes....

In the case of BHJ devices, the use of a TiO_x buffer layer inserted between the blend and the cathode gives considerable enhancement in the photocurrent attributed to the optical spacer effect of the TiO_x film [Kim et al., *Adv. Mater.*, 2006]. In the same order of idea, as discussed above, the introduction of an ultra-thin gold film (0.5 nm) (or a 3 nm thick MoO_3 film) at the anode/electron donor interface, allows strong solar cells improvement, whatever the transparent conductive oxide used, which demonstrates that well fitted buffer layer could allow in the new future to grow organic solar cells indium free.

The extent to which subtle interfacial buffer layers modification can change devices performance indicates that much room exists for improvement of organic solar cells.

Nowadays, two manufacturing techniques are mainly used, vacuum processing for the small organic molecules and wet processing for the polymers. In a first approach, wet processing, if transposable to simple ink jet process, has the advantage of low cost technique, however vacuum processing presents the advantages of easy fabrication of high quality thin films from highly purified materials, well controlled films thickness and therefore simplicity to construct complex multilayers structures. Even if, in a first approach, vacuum process appears more costly, its reproducibility can made it competitive for large scale massive production.

A cost factor of both cell families is oxygen and water protection. To day packaging is obtained using additive multilayer structures [Kim et al. *Appl. Phys. Lett.* (2009)], However, packaging techniques must be compatible with the substrate, which could be flexible, the active material and the final cost of the panel. One possible opportunity could be to use buffer layer which gives devices not only with high efficiency but also with long life structures.

The future of organic solar cells is bounded to the concept of "electronics everywhere", where light and flexible devices such as organic solar cells could play a major role in combination with thin films batteries...

10. References

T.D. Anthopoulos, B. Singh, N. Marjanovic, N.S. Sariciftci (2006). High performance n-channel organic field-effect transistors and ring oscillators based on C_{60} fullerene films, *Appl. Phys. Lett.*, 89, 213504.

N.R. Armstrong, C. Carter, C. Donley, A. Simmonds, P. Lee, M. Brumbach, B. Kippelen, B. Domercq, S. Yoo, (2003). Interface modification of ITO thin films: organic photovoltaic cells, *Thin Solid Films*, 445, 342-352.

M.A. Baldo, S.R. Forrest, (2001). Interface-limited injection in amorphous organic semiconductors, *Phys. Rev.* B 64, 085201.

A. J. Bard, L. R. Faulkner (1984). Electrochemical Methods. Fundamentals and Applications, Wiley, New York, p 634.

J.C Bernède, L. Cattin, M. Morsli, Y. Berredjem (2008). Ultra thin metal layer passivation of the transparent conductive anode in organic solar cells. *Solar Energy Materials and Solar Cells*, 92, 1508-1515.

J. C. Bernède, V. Jousseaume, M. A. Del Valle, F. R. Diaz (2001), Current Trends in polymer Sciences, From the organic electroluminescent diodes to the new organic photovoltaic cells, Vol 6, pp.135-155.

J.C. Bernede, Y. Berredjem, L. Cattin, M. Morsli (2008). Improvement of organic solar cells performances using a zinc oxide anode coated by an ultra   thin   metallic   layer. *Appl. Phys. Lett.,* 92, 083304.

Y. Berredjem, J.C. Bernede, S. Ouro Djobo, L. Cattin, M. Morsli, A. Boulmokh (2008). On the improvement of the efficiency of organic photovoltaic cells by the presence of an ultra-thin metal layer at the interface organic/ITO. *The European Physical Journal: Applied Physics*, 44, 223-228.

Y. Berredjem, N. Karst, A. Boulmokh, A. Drici, J.C. Bernede (2007) Optimisation of the interface "organic material/aluminium" of $CuPc/C_{60}$based photovoltaic cells, the *European Physical Journal: Applied Physics*, 40, 163-167.

Y. Berredjem, N. Karst, L. Cattin, A. Lakhdar-Toumi, A. Godoy, G. Soto, F. Diaz, M.A. del Valle, M. Morsli, A. Drici, A. Boulmkh, A.H. Gheid, A. Khelil, J.C. Bernède (2008). Plastic photovoltaic cells encapsulation, effect on the open circuit voltage. *Dyes and Pigments*, 78, 148-156.

J.C. Blakesley and N.C. Greenham (2009). Charge transfert at polymer-electrode interfaces: The effect of energetic disorder and thermal injection on band bending and open-circuit voltage disorder and thermal injection on band bending and open-circuit voltage, *J. Appl. Phys.*, 106, 034507.

C.J. Brabec., J. E. Shaheen, C. Winder, N.S. Sariciftci, P. Denk (2002). Effect of LiFÕmetal electrodes on the performance of plastic solar cells, *Appl. Phys. Lett.*, 80, 1.

S.S. Braun, W.R. Salaneck, M. Fahlman (2009). Energy-level alignment at organic/metal and organic/organic interfaces, *Adv. Mater.*, 21, 1450.

F. Brovelli, B.L. Rivas, J.C. Bernède, M. A. Del Valle, F. R. Diaz (2007). Electrochemical and optical studies of 1,4-diaminoanthraquine for solar cells applications, *Polym. Bull.*, 58, 521-527.

E.L. Bruner, N. Koch, A.R. Span, S.L. Bernasek, A. Kahn, J. Schwartz (2002). Controlling the Work Function of Indium Tin Oxide: Differentiating Dipolar from Local Surface Effects, *J. Am. Chem. Soc.*, 124, 3192-3193.

R. Cervini, X. C. Li, G. W. C. Spences, A. B. Holmes, S. C. Moratti, R. H. Friend. (1997). Electrochemical and optical studies of PPV derivatives and poly(aromatic oxadiazoles), *Synthetic Metals*, 84, 359-360.

M.Y. Chan, S.L. Lai, M.K. Fung, C.S. Lee, S.T. Lee, (2007). Doping-induced efficiency enhancement in organic photovoltaic devices, *Appl. Phys. Lett.*, 90, 023504.

J.A. Chaney, S.E. Koh, C.S. Dulcey, P.E. Pehrsson (2003). Surface chemistry of carbon removal from indium tin oxide by base and plasma treatment, with implications on hydroxyl termination, *Appl. Surf. Sci.*, 218, 259-267.

C-H. Chang, S.-A. Chen (2007). Effect of ionization potential change in poly(3,4-ethylenedioxythiophene):poly(styrenesulfonic acid) on the performance of polymer light emitting diodes due to its reaction with indium tin oxide, *Appl. Phys. Lett.*, 91, 103514.

F.-C. Chen, J.-L. Wu, C.-L. Lee, Y. Hong, C.-H. Kuo, M.H. Huang, (2009). Plasmonic-enhanced polymer photovoltaic devices incorporating solution-processable metal nanoparticles, Appl. Phys. Lett., 95, 013305.

L.L. Chen, W.L. Li, H.Z. Wei, B. Chu, B. Li (2006). Organic ultraviolet photovoltaic diodes based on copper phthalocyanine as an electron acceptor, *Sol. Energy Material*, 90, 1788-1796.

A.M. Cowley, S. M. Sze (1965). Surface States and Barrier Height of Metal-Semiconductor Systems, *J. Appl. Phys.*, 36, 3212.

X. Crispin. (2004). Interface dipole at organic/metal interfaces and organic solar cells, *Solar Energy Materials & Solar Cells* 83, 147-168.

S. Demmig, H. Langhals (1988). Leichtlösliche, lichtechte Perylen-Fluoreszenzfarbstoffe, *Chem. Ber.*, 121, 225-230.

S.H. Demtsu, J.R. Sites (2006). Effect of back-contact barrier on thin-film CdTe solar cells, *Thin Solid Films*, 510, 320-324.

C. Donley, D. Dunphy, D. Paine, C. Carter, K. Nebesny, P. Lee, D. Alloway, N.R. Amstrong (2002). Characterization of Indium−Tin Oxide Interfaces Using X-ray Photoelectron Spectroscopy and Redox Processes of a Chemisorbed Probe Molecule: Effect of Surface Pretreatment Conditions, Langmuir 18, 450-457.

J. Drechsel, B. Männig, D. Gebeyehu, M. Pfeiffer, K. Leo, H. Hoppe (2004). MIP-type organic solar cells incorporating phthalocyanine/fullerene mixed layers and doped wide-gap transport layers, *Organic electronics* 5 175-186.

D. Duche, P. Torchio, L. Escubas, F. Monestier, J.-J. Simon, F. Flory, G. Mathan, (2009). Improving light absorption in organic solar cells by plasmonic contribution, *Sol. Energy Mater. Sol. Cells*, 93, 1377-1382.

G. East, M. A. del Valle (2000). Easy to make Ag/AgCl reference electrode, *J. Chem. Educ.* 77(1) 97.

M. Fahlman, A. Crispin, X. Crispin, S.K.M. Henze, M.P. de Jong, W. Osikowicz, C. Tengstedt, W.R. Salaneck (2007). Electronic structure of hybrid interfaces for polymer-based electronics, *J. Phys.: Condens. Matter*, 19 183202.

Z.R. Hong, Z.H. Huang, X.T. Zeng, (2006). Investigation into effects of electron transporting materials on organic solar cells with copper phthalocyanine/C60 heterojunctions, *Chem. Phys. Lett.*, 425, 62-65.

H. Hoppe, N.S. Sariciftci (2004). Organic solar cells: An overview, *J. Mater. Res.*, 7, 1924-1945.

C.M. Hsu, W.-T. Wu, (2004). Improved characteristics of organic light-emitting devices by surface modification of nickel-doped indium tin oxide anode, *Appl. Phys. Lett.*, 85, 840.

J. Huang, J. Yu, H. Lin, Y. Jiang (2009). Detailed analysis of bathocuproine layer for organic solar cells based on copper phthalocyanine and C60, *J. Appl. Phys.*, 105, 073105.

H.C. Im, D.C. Choo, T.W. Kim, J.H. Kim, J.H. Seo, Y.K. Seo, Y.K. Kim (2007). Highly efficient organic light-emitting diodes fabricated utilizing nickel-oxide buffer layers between the anodes and the hole transport layers, *Thin Solid Films*, 515, 5099-5101.

H. Ishii, N. Hayashi, E. Ito, Y. Washizu, L. Sugi, Y. Kimura, M. Niwano, Y. Ouchi, K. Seki (2004). Kelvin probe study of band bending at organic semiconductor/metal interfaces: examination of Fermi level alignment, *Phys. Stat. Sol* (a) 201 1075_1094.

C. Jain, M. Willander, V. Kumar, (2007). Conducting Organic Materials and Devices, High-Detectivity Polymer Photodetectors with Spectral Response from 300 nm to 1450 nm *Semiconductors and Semimetals*, 81, 1-188.

X-Y. Jiang, Z.-L. Zhang, J. Cao, M.A. Khan, K.-ul-Haq, W.-Q. Zhu (2007). White OLED with high stability and low driving voltage based on a novel buffer layer MoOx, *J. Phys. D: Appl. Phys.*, 40, 5553-5557.

B. Johnev, M. Vogel, K. Fostoropulos, B. Mertesacker, M. Rusu, M.-C. Lux-Steiner, A. Weidinger (2005). Monolayer passivation of the transparent electrode in organic solar cells, *Thins Solid Films*, 488, 270-273.

B. Kang, L.W. Tan, S.R.P. Silva (2009). Ultraviolet-illuminated fluoropolymer indium–tin-oxide buffer layers for improved power conversion in organic photovoltaics, *Organic Electronics* 10 1178-1181.

K.S. Kang, H.K. Lim, K.J. Han, J. Kim (2008). Durability of PEDOT : PSS-pentacene Schottky diode, *J. Phys. D: Appl. Phys.*, 41, 012003.

N. Karst, J.C. Bernède (2006). On the improvement of the open circuit voltage of plastic solar cells by the presence of a thin aluminium oxide layer at the interface organic aluminium. *Physica Status Solidi (a)*, 203 R70-R72.

M.A. Katkova, V.A. Ilichev, A.N. Konev, M.A. Batenkin, L.L. Pestova, A.G. Vitukhnovsky, Bochkarev M.N. (2008). Modification of anode surface in organic light-emitting diodes by chalcogenes, *Appl. Surf. Sciences* 254 2216-2219.

S. Kato (2005). Designing Interfaces That Function to Facilitate Charge Injection in Organic Light-Emitting Diodes, J. Am. Chem. Soc., 127 11538-11539.

J.S. Kim, R.H. Friend, F. Cacialli (1999). Surface energy and polarity of treated indium–tin-oxide anodes for polymer light-emitting diodes studied by contact-angle measurements, *J. Appl. Phys.* 86, 2774.

J.Y. Kim, K. Lee, N.E. Coates, D. Moses, T.Q. Nguyen, M. Dante, A.J. Heeger, (2007) Efficient Tandem Polymer Solar Cells Fabricated by All-Solution Processing, *Science*, 317, 222

J.Y. Kim, S.H. Kim, H.-H. Lee, K. Lee, W. Lee, W. Ma, X. Gong, A.J. Heeger, (2006). New Architecture for High-Efficiency Polymer Photovoltaic Cells Using Solution-Based Titanium Oxide as an Optical Spacer, *Adv. Mater.*, 18 572-576.

N. Kim, W.J. Potscavage Jr., B. Domercq, B. Kippelen, S. Graham, (2009). A hybrid encapsulation method for organic electronics, *Appl. Phys. Lett.*, 94, 163308.

S.-S Kim, S.-I. Na, J. Jo, D.-Y. Kim, Y.-C. Nah, (2008). Plasmon enhanced perfomance of organic solar cells using electrodeposited Ag nanoparticles, *Appl. Phys. Lett.*, 93, 073307.

S.Y. Kim, K. H. Lee, B.D. Chin, J.-W. Yu (2009). Network structure organic photovoltaic devices prepared by electrochemical copolymerization, *Solar Energy Materials & Solar Cells* 93, 129-135.

Y. Kinoshita, T. Hasobe, H. Murata (2007). Control of open-circuit voltage in organic photovoltaic cells by inserting an ultrathin metal-phthalocyanine layer, Appl. Phys. Lett., 91, 083518.

Th. Kugler, W. R. Salaneck, H. Rast, A. B. Holmes (1999). Polymer band alignment at the interface with indium tin oxide: consequences for light emitting devices, *Chem. Phys. Lett.*, 310, 391-386.

B. Kouskoussa, M. Morsli. K. Benchouk, G. Louarn, L. Cattin, A. Khelil, J.C. Bernede, (2009). On the improvement of the anode/organic material interface in organic solar cells by the presence of an ultra-thin gold layer. *Physica Status Solidi (a)* 206, 311-315.

Th. Kugler, A. Johansson, I. Dalsegg, U. Gelius, W.R. Salanek (1997). Electronic and chemical structure of conjugated polymer surfaces and interfaces: applications in polymer-based light-emitting devices, *Synthetic Metals*, 91, 143-146.

Th. Kugler, W. R. Salaneck, H. Rast, A. B. Holmes (1999). Polymer band alignment at the interface with indium tin oxide: consequences for light emitting devices, *Chem. Phys. Lett.* 310, 391-396.

A. Latef, J. C. Bernède (1991). Study of the thin film interface Aluminium –Tellurium, *Phys. Stat. Sol. (a)*, 124, 243.

C.S. Lee, J.X. Tang, Y.C. Zhou, S.T. Lee, Interface dipole et metal-organic interfaces: contribution of metal induced interface states (2009). *Appl. Phys. Lett.* 94 113304.

C.N. Li, C.Y. Kwong, A.B. Djurisic, P.T. Lai, P.C. Chui, W.K. Chan, S.Y. Liu (2005). Improved performance of OLEDs with ITO surface treatments, *Thins Solid Films* 477 57-62.

Y. Li, Y. Cao, D. Wang, G. Yu, A. S. Heeger (1999). Electrochemical properties of luminescent polymers and polymer light-emitting electrochemical cells, *Synthetic Metals*, 99, 243-248.

S. G. Liu, G. Sui, R. A. Cormier, R. M. Leblanc, B. A. Gregg (2002). Self-Organizing Liquid Crystal Perylene Diimide Thin Films: Spectroscopy, Crystallinity, and Molecular Orientation, *J. Phys. Chem. B*, 106, 1307-1315.

Lord Kelvin (1898). *Philos. Mag.*, 46, 82.

H-T. Lu, M. Yokoyama (2004). Plasma preparation on indium-tin-oxide anode surface for organic light emitting diodes, *Journ. Crys. Growth*, 260, 186-190.

D. Lüthi, , M. Le Floch, B. Bereiter, T. Blunier, J.-M. Barnola, U. Siegenthaler, D. Raynaud, J. Jouzel, H. Fischer, K. Kawamura, and T.F. Stocker (2008). High-resolution Carbon dioxide concentration record 650,000–800,000 years before present, Nature. 453 379-382.

T. Matsushima, Y. Kinoshita, H. Murata (2007). Formation of Ohmic hole injection by inserting an ultrathin layer of molybdenum trioxide between indium tin oxide and organic hole-transporting layers, *Appl. Phys. Lett.*, 91, 253504.

A.G. Milnes, D.L. Feucht 1972. Heterojunctions and metal semi-conductor junctions. Academic Press Editor, New York,

A.J. Morfa,. K.L. Rowlen, T.H. Reilly III, M.J. Romero, J. van de Lagemaat, (2008). Plasmon-enhanced solar energy conversion in organic bulk heterojunction photovoltaics, *Appl. Phys. Lett.*, 92, 013504.

M. Niggemann, M. Riede, A. Gombert, K. Leo (2008). Light trapping in organic solar cells, *Phys. Stat. Sol. (a,)* 205, 2862-2874.

T. Nishi, K. Kanai, Y. Ouchi, M.R. Willis, K. Seki (2005). Evidence for the atmospheric *p*-type doping of titanyl phthalocyanine thin film by oxygen observed as the change of interfacial electronic structure, *Chem. Phys. Lett.*, 414, 479-482.

F. Nüesch, L.J. Rothberg, E.W. Forsythe, Q. Toan Le, Y. Gao, (1999). A photoelectron spectroscopy study on the indium tin oxide treatment by acids and bases, *Appl. Phys. Lett.*, 74, 880.

F. Padinger, R.S. Rittenberger, N.S. Sariciftci (2003). Effects of Postproduction Treatment on Plastic Solar Cells, *Adv. Funct. Mater.*, 13, 85-88.

S. Y. Park, Y. H. Noh, H. H. Lee (2006). Introduction of an interlayer between metal and semiconductor for organic thin-film transistors, *Appl. Phys. Lett.*, 88, 113503.

P. Peumans, V. Bulovic, S.R. Forrest (2000). Efficient photon harvesting at high optical intensities in ultrathin organic double-heterostructure photovoltaic diodes, *Appl. Phys. Lett.* 76, 2650.

P. Peumans, S.R. Forrest (2001). Very-high-efficiency double-heterostructure copper phthalocyanine/C_{60} photovoltaic cells, *Appl. Phys. Lett.* 79, 126-128

M. Pfeiffer, K. Leo, N. Karl (1996). Fermi level determination in organic thin films by the Kelvin probe method, *J. Appl. Phys.* 80, 6880,

K.L. Purvis, G. Lu, J. Schwartz, S.L. Bernasek (2000). Surface Characterization and Modification of Indium Tin Oxide in Ultrahigh Vacuum, *J. Am. Chem. Soc.*, 122, 1808-1809.

C. Qiu, Z. Xie, H. Chen, M. Wong, H.S. Kwok (2003). Comparative study of metal or oxide capped indium–tin oxide anodes for organic light-emitting diodes, *J. Appl. Phys.*, 93, 3253.

B.P. Rand, D.P. Burk, S.R. Forrest (2007). Offset energies at organic semiconductor heterojunctions and their influence on the open-circuit voltage of thin-film solar cells, *Phys. Rev.* B 75, 115327,

B.P. Rand, J. Li, J. Xue, R.J. Holmes, M.E. Thompson, S.R. Forrest (2005). Organic double-heterostructure photovoltaic cells employing thick tris(acetylacto)ruthenium(III) exciton-blocking layers, *Adv. Mater.*, 17, 2714-2718.

C.A. Rozzi, F. Mangi, F. Parmigiani (2003). *Ab initio* Fermi surface and conduction-band calculations in oxygen-reduced MoO_3, *Phys. Rev.*, B 68, 075106.

N.S. Sariciftci, D. Braun, C. Zhang, V. I. Srdanov, A. S. Heeger, G. Stucky, F. Wuld (1993). Semiconducting polymer-buckminsterfullerene heterojunctions: Diodes, photodiodes, and photovoltaic cells, *Appl. Phys. Lett.*, 62, 585.

N. S. Sariciftci, L. Smilowitz, A. J. Heeger, and F. Wudl, (1992). Photoinduced Electron Transfer from a Conducting Polymer to Buckminsterfullerene, *Science*, 258, 5087, 1474-1476.

J.C. Scott, J.H. Kaufman, P.J. Brock, R. DiPietro, J. Salem, J.A. Goitia (1996). Degradation and failure of MEH-PPV light-emitting diodes, *J. Appl. Phys.*, 79, 2745.

K. Seki, E. Ito, H. Ishii (1997). Energy level alignment at organic/metal interfaces studied by UV photoemission, *Synthetic Metals*, 91, 137-142,

S.E. Shaheen, C.J. Brabec, N.S. Sariciftci, F. Padinger, T. Fromherz, J.C. Hummelen, (2001). 2.5% efficient organic plastic solar cells, *Appl. Phys. Lett.*, 78, 841-843.

A. Sharma, P. J. Hotchkiss, S.R. Marder, B. Kippelen, (2009). Tailoring the work function of indium tin oxide electrodes in electrophosphorescent organic light-emitting diodes, *J. Appl. Phys.*, 105, 084507.

S.W. Shi, D.G. Ma, J.B. Peng (2007). Effect of NaCl buffer layer on the performance of organic light-emitting devices (OLEDs), Eur. Phys. J. Appl. Phys., 40, 141-144.

V.P. Singh, B. Parthasarathy, R.S. Singh, A Aguilera, J. Antony, M. Payne (2006). Characterization of high-photovoltage CuPc-based solar cell structures *Solar Energy Materials and Solar Cells*, 90, 798-812.

V.P. Singh, R.S. Singh, B. Parthasarathy, A Aguilera, J. Antony, M. Payne (2005). Copper-phthalocyanine-based organic solar cells with high open-circuit voltage, *Appl. Phys. Lett.*, 86, 082106,

Q.L. Song, F.Y. Li, H. Yang, H.R. Wu, X.Z. Wang, W. Zhou, J.M. Zhao, X.M. Ding, C.H. Huang, X.Y. Hou (2005). Small-molecule organic solar cells with improved stability, Chem. Phys. Lett., 416, 42.

P. Sullivan, T.S. Jones, A.J. Ferguson (2007). Structural templating as a route to improved photovoltaic performance in copper phthalocyanine/fullerene (C60) heterojunctions, *Appl. Phys. Lett.*, 91, 233114.

S.M. Sze 1981"Physics of Semiconductor Devices" 2nd Edition, John Wiley Editor, New York.

Y. Tanaka, K. Kanai, Y. Ouchi, K. Seki (2009). Role of interfacial dipole layer for energy-level alignment at organic/metal interfaces, *Organic Electronics*, 10, 990-993.

C.W. Tang (1986). Two-layer organic photovoltaic cell, *Appl. Phys. Lett.*, 48, 183.

C. Tengstedt, W. Osikowioz, W.R. Salaneck, I.D. Parker, C.-H. Hsu, M. Fahlman (2006). Fermi-level pinning at conjugated polymer interfaces, *Appl. Phys. Lett.*, 88, 053502.

B.C Thompson., J.M.J. Fréchet (2008). Polymer-Fullerene Composite Solar Cells, *Angew. Chem. Int. Ed.*, 47, 58-77.

J. Van de Lagemaat, T.M. Barnes, G. Rumbles, S. E. Shaheen, T.J. Coutts, C. Weeks, I. Levitsky, J. Peltola, P. Glatkowski (2006). Organic solar cells with carbon nanotubes replacing In2O3:Sn as the transparent electrode, *Appl. Phys. Lett.*, 88, 233503.

J. Xue, B.P. Rand, S. Uchida, S.R. Forrest, (2005). Mixed donor-acceptor molecular heterojunctions for photovoltaic applications. II. Device performance, *J. Appl. Phys.*, 98, 124903.

I. Yoo, M. Lee, C. Lee, D.-W. Kim, I.S. Moon, D.-H. Hwang (2005). The effect of a buffer layer on the photovoltaic properties of solar cells with P3OT:fullerene composites, Synthetic Metals, 153, 97-100.

H. You, Y. Dai, Z. Zhang, D. Ma (2007). Improved performances of organic light-emitting diodes with metal oxide as anode buffer, J. Appl. Phys., 101, 26105.

X-H. Zhang, B. Domercq, B. Kippelen (2007). High-performance and electrically stable C60 organic field-effect transistors, *Appl. Phys. Lett.* 91 092114.

D. Zhang, Y. Li, Guohui Zhang, Y. Gao, L. Duan, L. Wang, Y. Qiu Lithium cobalt oxide as electron injection material for high performance organic light-emitting diodes (2008). *Appl. Phys. Lett.*, 92, 073301.

Z.Y. Zhong, Y.D. Jiang (2006). Surface treatments of indium-tin oxide substrates for polymer electroluminescent devices, *Phys. Stat. Sol., (a)*, 203, 3882-3892.

Y. Zhou, Y. Wang, W. Wu, H. Wang, l. Han, W. Tian, H. Bässler (2007). Spectrally dependent photocurrent generation in aggregated MEH-PPV:PPDI donor–acceptor blends, *Sol. Energy Mat. Solar cells*, 91, 1842-1848.

B. Zimmermann, M. Glatthaar, M. Niggermann, M. Riede, A. Hinsch (2005). Electroabsorption studies of organic bulk-heterojunction solar cells, *Thin Solid Films*, 493, 170-174.

W.A. Zisman. (1932). A new method of measuring contact potential differences in metals, Rev. Sci. Instrum. 3, 367-370.

5

Efficient Silicon Solar Cells Fabricated with a Low Cost Spray Technique

Oleksandr Malik and F. Javier De la Hidalga-W.
National Institute for Astrophysics, Optics and Electronics (INAOE)
Mexico

1. Introduction

Since the 1960s, studies of transparent and highly conducting semiconductor metal oxide films, such as tin-doped indium oxide (ITO) and fluorine-doped tin oxide (FTO), have attracted the interest of many researches due to their wide applicability in both industry and research, such as transparent heat reflecting films, gas sensors, protective coatings, and heterojunction solar cells. Excellent reviews of the subject can be found in the literature (Hamberg & Granquist, 1986, Granquist, 1993, Hartnagel et al, 1995, Dawar & Joshi, 2004). Several methods such as chemical vapour deposition, vacuum evaporation, sputtering techniques, magnetron sputtering, ion implantation, ion-beam sputtering, and spraying techniques have been tried out to fabricate coatings for the oxide materials (Dawar & Joshi, 2004). This chapter is intended to be a comprehensive review of the original results in the field of fabrication of ITO and FTO films by spray pyrolysis technique, as well as their applications for the fabrication of efficient silicon (Si) monocrystalline solar cells and modules with a low fabrication cost. Since the end of the 1970s, such solar cells have attracted the attention of the scientific community because of their fabrication simplicity (DuBow et al., 1976, Manifacier & Szepessy, 1977, Feng at al., 1979, Malik et all., 1979, 1980). It has been shown that those metal oxide films deposited on the silicon surface form heterojunctions, which at a first approximation, can be considered as Schottky "transparent" metal-semiconductor (MS) contacts due to the degeneracy of the electron gas occurring in the n-type highly conducting transparent metal oxide films, such as ITO and FTO. Depending on the method used for the deposition of the films and the type of the silicon substrate conductivity, such contact can work as a rectifier and the surface-barrier solar cell can be designed based on it. Thus, films deposited in vacuum on the p-silicon surface form rectifying contacts (DuBow et al., 1976), while electron-beam (Feng at al., 1979), whereas sprayed films deposited on n-type silicon form rectifying contacts (Manifacier & Szepessy, 1977, Malik et all., 1979). This phenomenon occurs because of the difference in electron affinity of the ITO and FTO films fabricated with the methods mentioned above. Further studies have shown that ITO/Si and FTO/Si solar cells operate almost always as majority-carriers MS or metal-insulator-semiconductor (MIS) diodes. The formation of a very thin SiO_x layer between the metal oxide film and the silicon leads to an increasing efficiency of the solar cells based on MS contacts (Ashok et al., 1980). However, if the potential barrier at the silicon surface is very high, an inversion p-n layer is formed at the silicon surface. In this case the properties of these solar cells will be similar to those of solar cells based on metallurgical p-n junctions obtained by diffusion of dopants in a silicon substrate. This was demonstrated with ITO/Si structures fabricated by sputtering of the ITO film on a p-type silicon substrate (DuBow et al., 1976), where the solar cells operated as minority-carriers (or p-n) diodes. On the other hand, numerous published works have shown that solar cells fabricated on n-type silicon operate only as majority (MS or MIS) diodes (Nagatomo et al., 1979, 1982). Nevertheless, in this work we show that combining a special treatment of the silicon surface with the electro-physical properties of the spray deposited films allows for the fabrication of inversion p-n solar cells based on n-type silicon substrates.

2. Fabrication of ITO and FTO films by spray pyrolysis

The spray pyrolysis technique was used for the deposition of thin ITO and FTO films on glass and sapphire substrates in order to investigate their structural, electric, and optical properties. A 10 Ω-cm n-type (100)-oriented silicon substrate, whose surface was chemically cleaned and specially treated, was used for the fabrication of the solar cells. The glass apparatus (atomizer) was designed in such a way that small-size droplets were obtained. The substrates were mounted on a heater covered with a carbon disk in order to assure a uniform temperature, and the spraying was conducted using compressed air. Periodical cycles of the deposition with durations of 1 sec and intervals of 5 sec were employed to prevent a rapid substrate cooling. Deposition rate was high, about 200 nm/min.

2.1 Deposition of ITO films
For the ITO films deposition, 13.5 mg of $InCl_3$ were dissolved in 170 ml of 1:1 water and ethylic alcohol mixture, with an addition of 5ml of HCl. Different ratios of Sn/In achieved in the ITO films were controlled by adding in the solution a calculated amount of tin chloride ($SnCl_4*5H_2O$). The substrate temperature was controlled with a thermocouple at a value of 480 ± 5^0C. The optimal distance from the atomizer to the substrate and the compressed air pressure were 25 cm and 1.4 kg/cm^2, respectively. We obtained a high deposition rate of about 200 nm/min.

2.2 Deposition of FTO films
The precursors for the deposition of the FTO film were prepared based on 0.2 M alcoholic solutions of $SnCl_4*5H_2O$ with different content of NH_4F for obtaining different F/Sn ratio in the films. The remaining deposition parameters were the same as those used for depositing the ITO films.

2.3 Characterization equipment and methods
The film thickness was measured with an Alpha Step 200 profilometer. The electrical resistivity, Hall mobility and carrier concentration were measured at room temperature using the van der Pauw method. Hall effect parameters were measured using a magnetic field of 0.25 Tesla. The optical transmission spectra were obtained with a spectrophotometer. The structural characterization was carried out with an X-ray diffractometer operating in the Bragg-Brentano Θ-Θ geometry with Cu K$_\alpha$ radiation. A JSPM 5200 atomic force microscope was used to study the film surfaces. The chemical composition of the films was determined using an UHV system of VG Microtech ESCA2000 Multilab with an Al- K$_\alpha$ X-ray source (1486.6 eV) and a CLAM4 MCD analyzer.

3. Brief description of the film properties

3.1 Tin-doped indium oxide (ITO) films

The X-ray diffraction (XRD) measurements shown in Figure 1 indicate that all deposited ITO films, with thickness 160-200 nm and fabricated from the chemical solutions with different Sn/In ratio, present a cubic bixebyte structure in a polycrystalline configuration with a (400) preferential grain orientation.

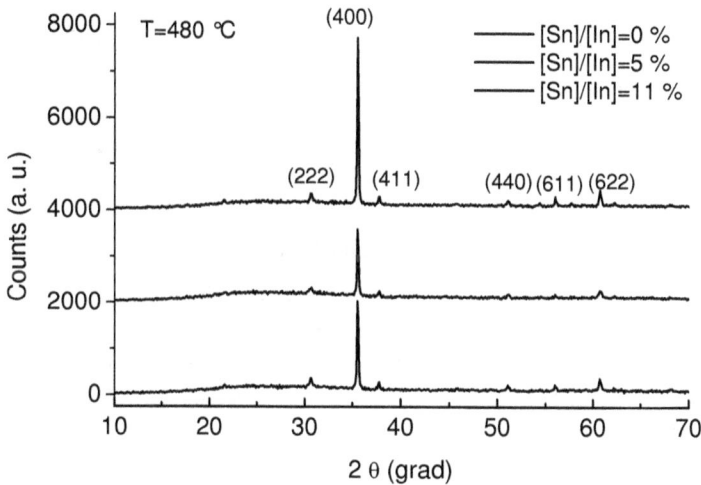

Fig. 1. XRD spectra of the ITO films fabricated from precursors with different Sn/In ratio

The average size of the grains, 30-50 nm, was determined using the classical Debye-Scherrer formula from the half-wave of the (400) reflections of the XRD patterns

A surface roughness about 30 nm was determined from images of the films surfaces obtained with the atomic force microscope (Figure 2).

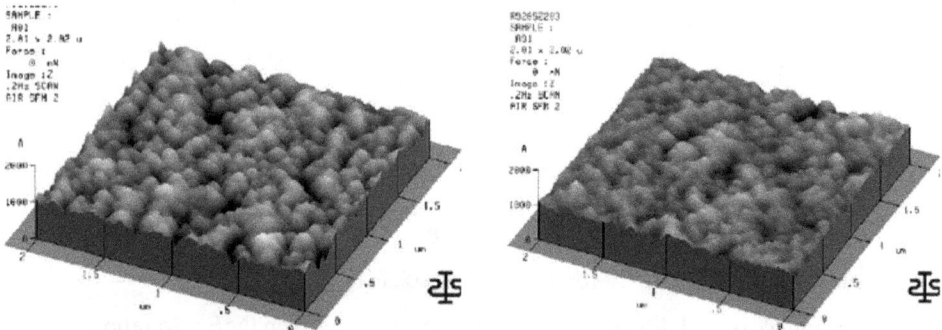

Fig. 2. AFM images of the In_2O_3 film (left) and the ITO film with 5% Sn/In (right)

Figures 3 and 4 show the dependence of electric parameters of the spray deposited ITO film on the ratio Sn/In. The sheet resistance R_s shown in Figure 3 presents a minimum of 12 Ω/□ the films prepared from the solution with a 5% Sn/In ratio.

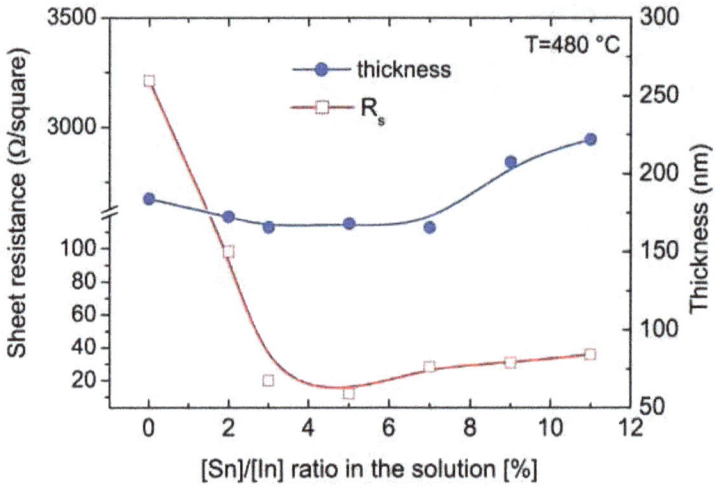

Fig. 3. The sheet resistance as a function of the Sn/In ratio in the precursor used for the film deposition. The thicknesses of the films are also shown

The minimal value of resistivity obtained for the films deposited for the solution with 5% Sn/In ratio is 2×10^{-4} Ω-cm. The variation of mobility and carrier concentration as a function of the Sn/In ratio are shown in Figure 4.

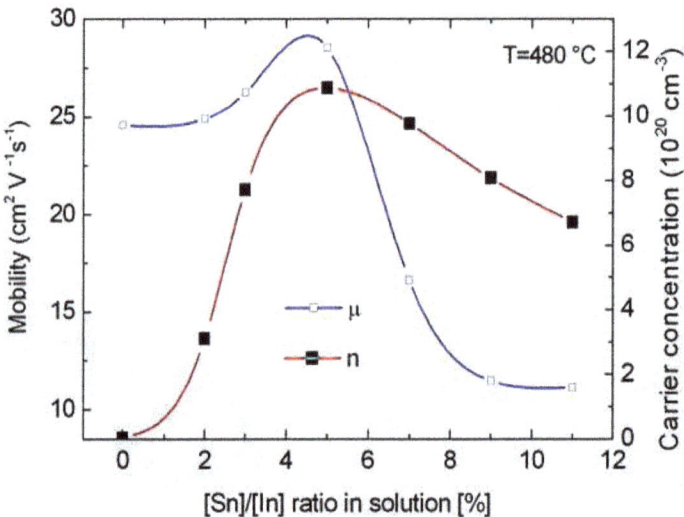

Fig. 4. Dependence of mobility (μ) and carrier concentration (n) on the Sn/In ratio

Figure 5 shows the optical transmission spectra for the ITO films spray-deposited on a sapphire substrate as a function of the wavelength for solutions with different Sn/In contents.

The use sapphire substrates allow for determining the optical energy gap of the ITO films by extrapolating the linear part of α^2(hv) curves to $\alpha^2=0$, where α is the absorption coefficient.

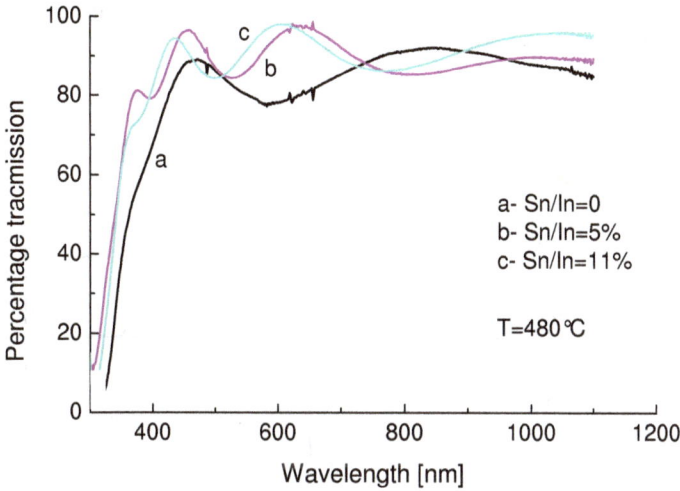

Fig. 5. Optical transmission spectra for the ITO films spray-deposited for different precursors as a function of the wavelength

The optical gap increases with the carrier concentration, corresponding to the well known Burstein-Moss shift. For the Ito films fabricated using the solution with a 5% Sn/In ratio this shift is 0.48 eV, and the optical gap is 4.2 ± 0.1 eV. Such high value for the optical gap offers transparency in the far ultraviolet range, which is important for the application of these films in solar cells.

Because of the opposite dependence of the conductivity (σ) and transmission (T) on the thickness (t) of the ITO, both parameters need to be optimized.

A comparison of the performance for different films is possible using the $\phi_{TC}=T^{10}/R_s=\sigma t$ $\exp(-10\alpha t)$ figure of merit (Haacke, 1976). Table 1 compares the values of ϕ_{TC} for the spray deposited ITO films reported in this work with some results obtained by other authors using different deposition techniques.

Process	R_s, $\Omega/\square$	T (%)	ϕ_{TC}, $(\Omega^{-1}) \times 10^{-3}$	Author
spray	26.0	90	13.4	Gouskov, 1983.
spray	9.34	85	21.0	Vasu et al., 1990
spray	10.0	90	34.9	Manifacier, 1981
spray	4.4	85	44.7	Saxena, 1984
sputtering	12.5	95	47.9	Theuwissen, 1984
evaporation	25.0	98	32.6	Nath, 1980
spray	12.0	93.7	43.5	Present work

Table 1. Comparison of the values of ϕ_{TC} for ITO films

3.2 Fluorine-doped tin oxide (FTO) films

The X-ray diffraction (XRD) measurements indicate that all the spray-deposited FTO films present a tetragonal rutile structure in a polycrystalline configuration with a (200)

preferential grain orientation. The XRD spectra of the FTO films fabricated using precursors with different F/Sn ratios are shown in Figure 6.

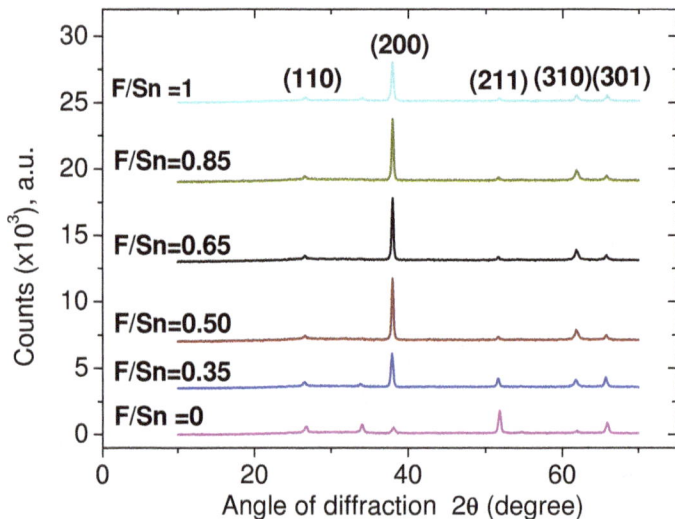

Fig. 6. The XRD spectra for the FTO films fabricated using precursors with different F/Sn ratio

The surface morphology of the films fabricated using precursors with different F/Sn ratio, and obtained with a scanning electron microscopy (SEM), is shown in Figure 7.

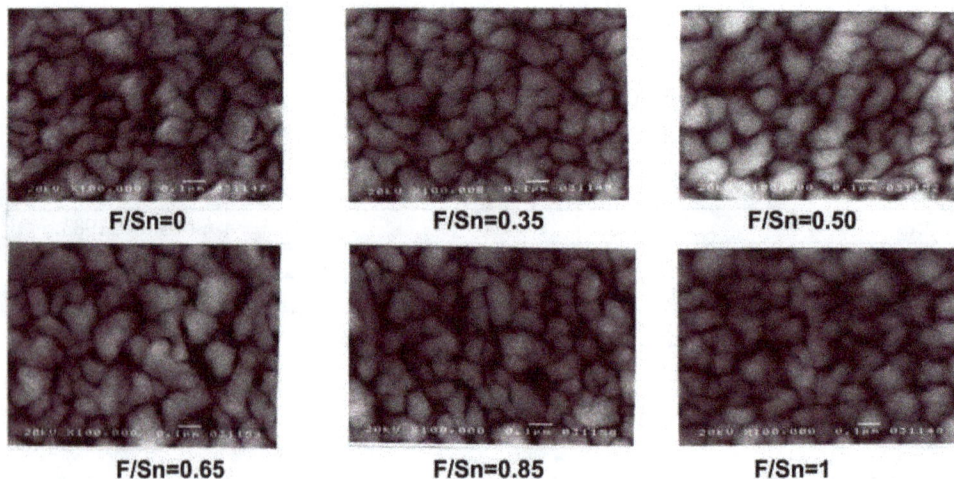

Fig. 7. The surface morphology obtained with a SEM for the films fabricated using precursors with different F/Sn ratios

The dependence of the average value of the grain size on the F/Sn ratio shows a maximum (~ 40 nm) for the films prepared using a precursor with F/Sn=0.5. The roughness variation

obtained with atomic force microscope for the FTO film fabricated using solutions with different F/Sn ratios presents a minimum of 8-9 nm at the F/Sn=0.5 ratio.
Figure 8 shows that the electrical characteristics also present some peculiarities for the films prepared using a precursor with this F/Sn ratio.

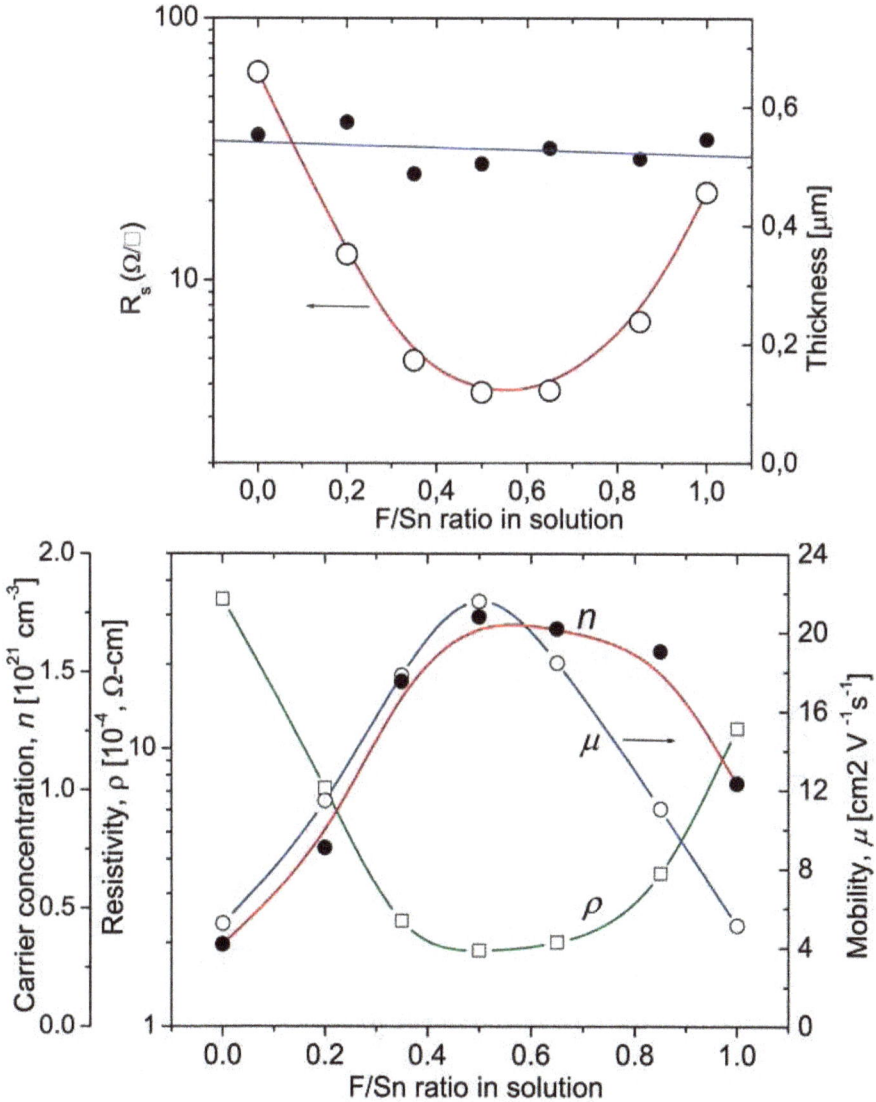

Fig. 8. Variation of the sheet resistance (above graph), resistivity (ρ), mobility (μ) and carrier concentration (n) (below graph) for the FTO films fabricated using precursors with different F/Sn ratios. The thicknesses of the films are also shown

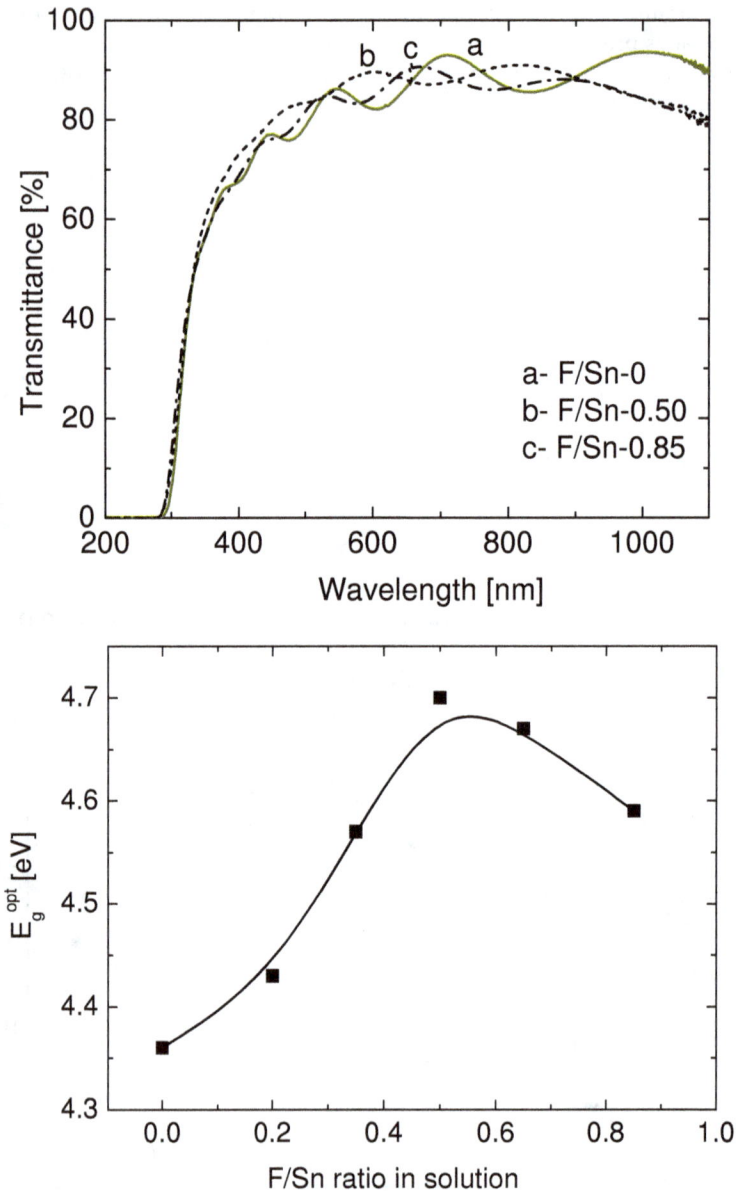

Fig. 9. Optical transmission (above graph) and dependence of the optical gap (below graph) for the FTO films fabricated using solutions with different F/Sn contents and spray-deposited on a glass substrate as a function of the wavelength

The optical energy gap (Fig. 9) was determined from the analysis of the absorption spectra for the films deposited on the sapphire substrate. The Burstein-Moss shift presents a maximum value of 0.6 eV for the films fabricated using the precursor with F/Sn =0.5, which also corresponds to the highest electron concentration (1.8×10^{21} cm^{-3}). Figure 10 shows the $\Phi=T^{10}/R_s$ figure of merit for the FTO films reported in this work.

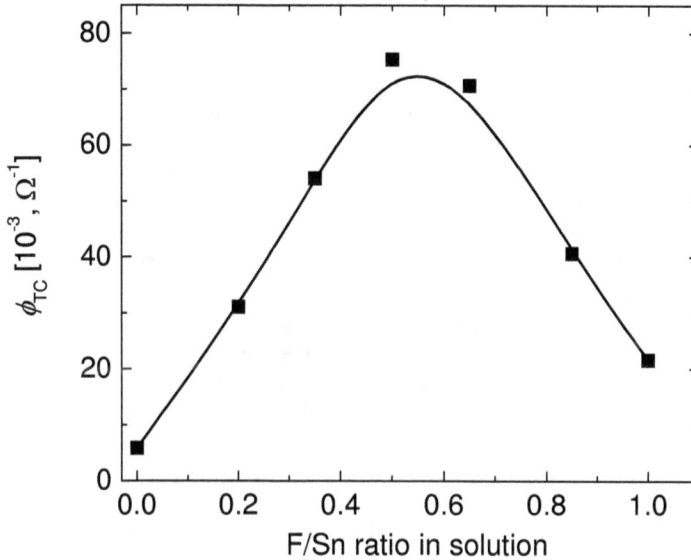

Fig. 10. Variation of the figure of merit $\Phi=T^{10}/R_s$ versus the F/Sn ratio used in the solution for the FTO films reported in this work

The value we obtained for this figure of merit was Φ =75×10^{-3} Ω^{-1} for the films prepared using a precursor with F/Sn =0.5; this is more than twice the value (Φ =35×10^{-3} Ω^{-1}) reported in the literature (Moholkar et al., 2007) for spray deposited FTO films.

4. Solar cells based on ITO/n-Si heterojunctions

4.1 Physical model of the solar cells
When the ITO (or FTO) film is deposited on the silicon surface, a metal-semiconductor contact-like is formed due to the metallic electric properties of the degenerated metal oxide. Ideally, the barrier height (φ_b) formed between the metal and the n-type semiconductor is determined by the difference between the metal (or in our case the metal oxide) work function (φ_M) and the electron affinity (χ_s) in the semiconductor. Actually, the surface states present in the interface pin the Fermi level, which makes the barrier height less sensitive to the metal work function (Sze, 2007). The surface has to experiment a reconstruction due to the discontinuity of the lattice atoms on the surface. Each surface atom present a dangling bond and shares a dimer bond with its neighbor atom, thus giving place to surface states inside the Si band gap (Trmop, 1985).

Recently, it has been shown that the barrier height in a metal-silicon junction can take an almost ideal value if the n-Si surface is passivated with sulfur (Song, 2008). Also the open-circuit voltage of an Al/ultrathin SiO$_2$/n-Si solar cell (Fujiwara, 2003) was improved when the silicon surface was passivated by a cyanide treatment.

In this chapter we will discuss the properties of the ITO/n-Si solar cells presenting extremely high values of the potential barrier at the silicon interface obtained by passivating the surface with a hydrogen-peroxide solution.

If the ITO film is deposited on cleared n-type silicon, the barrier height not exceeds 0.76 eV. For this value of the barrier height, the ITO/nSi heterojunctions fabricated on silicon substrates with a resistivity of a few Ω-cm, operate as majority carrier devices, whose characteristics are well described by the Schottky theory. Usually, such type of devices present a high value for the dark current originated by the thermo-ionic mechanism, and the open circuit voltage for these structures designed as solar cells shows a sufficiently low value. The introduction of a very thin (~ 2 nm) intermediate SiO_x layer (Feng, 1979) decreases the dark current and increases the open-circuit voltage. However, the use of this approach to improve the characteristics of the surface-barrier solar cells requires a simultaneous and careful control of the intermediate oxide thickness. Furthermore, the thermal grown intermediate SiO_x layer always presents a positive fixed charge located at the SiO_x/Si interface, which decreases the barrier height in the case of n-type silicon.

Using known data for the work function of ITO films deposited by spray pyrolysis, whose average value is reported as 5.0 eV (Nakasa et al., 2005, Fukano, 2005), and the electron affinity of silicon as 4.05 eV, the ideal barrier height between ITO and n-type silicon is 0.95 eV according to the Mott-Schottky theory. After a treatment of the n-type silicon surface in the hydrogen-peroxide (H_2O_2) solution with a controlled temperature (60 ^{0}C) during 10 minutes, a barrier height of 0.9 eV was obtained with capacitance-voltage measurements. This value exceeds by 0.14 eV the barrier height obtained after the deposition of the ITO film on the silicon surface cleaned in HF without the treatment in an H_2O_2 solution.

It is worth discussing the possible reason for this increment of the barrier height after the treatment of the silicon surface, as well as the operation of the ITO/n-Si junctions with an extremely high barrier height. Obviously, a junction with such barrier height fabricated on the silicon substrates with moderate resistivity could behave as p-n junctions, in which a surface p-layer is induced by the high surface band bending.

Such situation was obtained (Shewchun, 1980) in solar cells ITO/ultrathin SiO_x/p-Si structures. However, in this case the inversion of the conductivity type of the p-Si at the surface was caused by other factors, such as the low work function of the sputtered ITO film and the presence of positive charge at the SiO_x/p-Si interface.

What is the physical reason for the increment of the barrier height in the ITO/n-Si heterojunctions after the treatment of the silicon substrate in heated 30% H_2O_2 solutions? It has been shown (Verhaverbeke, 1997) that the treatment of the silicon in H_2O_2 leads to the growth of oxide on the silicon surface. The analysis shows that the main oxidant responsible for this oxide growth is the peroxide anion, HO_2^-. It was also found that the oxide thickness is limited to a value around 0.8-1.0 nm due to the presence of localized negative charge (HO_2^-) at the silicon surface. From this point of view the HO_2^- at the silicon surface can play a double role. First, these ions can form a chemical composition with the silicon atoms having dangling bounds in the surface. This can be thought as a passivation of the silicon surface, which leads to an increment of the potential barrier during the formation of the ITO/Sl heterostructure. On the other hand, the negative charge of these ions can produce a band-bending (φ_s) at the silicon surface due to an outflow of electrons under the influence of the electrostatic force. Under such conditions, the electron affinity (χ_s) of the silicon at the surface will be lower than that at the bulk by $\Delta\chi = \chi_s - \varphi_s$. The presence of a depletion layer at the silicon surface plays an important role for the formation of the potential barrier during the deposition of the ITO film. The barrier will prevent an electron flow from the silicon to the ITO film. The surface barrier between the ITO and the silicon will be formed by the flow of valence electrons from the silicon valence band into the ITO film, creating a hole excess at the silicon surface. Taking into account the initial band-bending at the silicon surface, the formation of an inversion layer is possible. As it was already mentioned, the experimentally determined barrier height at the ITO/Si interface is 0.9 eV. Schematically, the energy diagram of the ITO/n-Si heterojunction is shown in Figure 11.

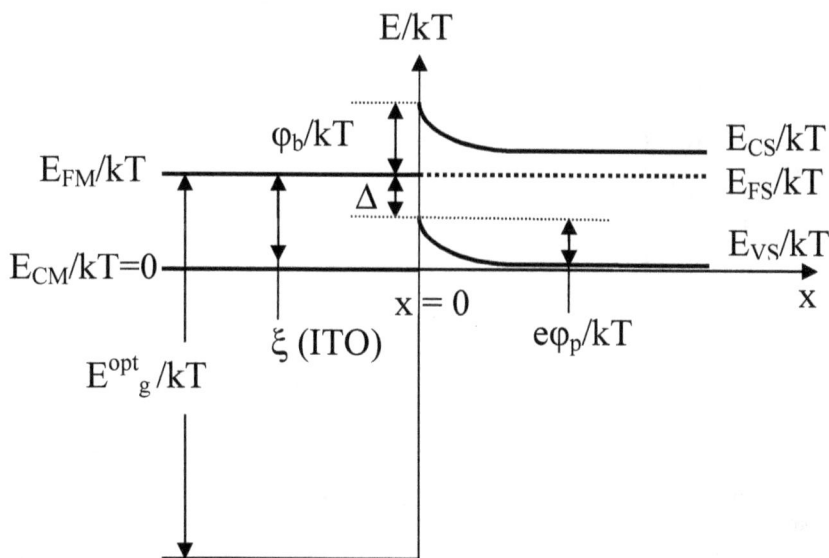

Fig. 11. Energy diagram (in kT units) of the heavy doped ITO/n-Si heterojunction

For sake of simplicity, we do not show the very thin (around 1 nm) intermediate SiO_x layer present between the ITO film and the silicon, because at this thickness it does not present any effect on the electro-physical characteristics of the heterojunction. Since the heavily doped ITO film is a degenerated semiconductor, in which the Fermi level lies above the minimum of the conduction band, we can consider this ITO film as a "transparent metal." The inversion layer at the silicon interface appears when the barrier height φ_b is higher than one-half of the Si energy gap. If such inversion p-n junction were connected in a circuit, which source of holes would be present in order to form an inversion p-layer that complicates the current flow across the forward-biased structure working as a solar cell? To answer this question we calculated the number of empty energy states in the conduction band of a heavy doped ITO, which are available to accept the electrons transferred from the top of the silicon valence band located at a distance Δ below the Fermi level (Malik et al., 2006). The probability that an energy state E below the Fermi level E_{FM} in the degenerated ITO is empty was calculated using the Fermi-Dirac distribution. Using a barrier height φ_n=0.9 eV, Δ=0.3 eV, and three different values for (E_{FM}-E_{CM}), which is the distance between the Fermi level an the conducting band of the ITO. This characterizes the degree of degeneration of the ITO film. The calculated number of empty states available to accept the electrons from the silicon valence band forming the additional amount of the holes is shown in Figure 12 as triangles. For comparison the number of empty states in the case of a gold/silicon contact with the same barrier height is also shown. For such calculations, the difference between the effective mass of electrons in the ITO and that in gold has been taken into account.

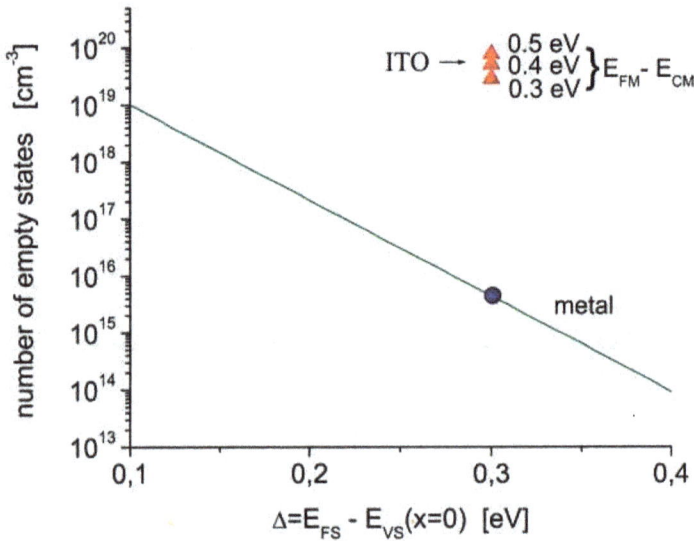

Fig. 12. Calculated number of empty states available to accept the electrons from the silicon valence band (Malik et al., 2006)

From the discussion presented above, and the amount of the calculated number of empty states in the ITO, leads to the important conclusion that a heavy doped ITO layer serves as an efficient source of holes necessary to form the inversion p-layer in the ITO/n-Si structures.

4.2 Evidence of the inversion in the type conductivity in the ITO/n-Si heterostructures

Based on the barrier height (0.9 eV) obtained from the measured C-V characteristics for the ITO/n-Si heterostructures on 10 Ω-cm monocrystalline silicon, one can discuss about the physical nature of such heterojunctions. Because the barrier height exceeds one half of the silicon band gap, the formation of an inversion p-layer at the silicon surface is obvious from the band diagram. To avoid any speculations on this issue and in order to present a clear evidence for the existence of a minority (hole) carrier transport in these heterojunctions, a bipolar transistor structure was fabricated on a 10 Ω-cm monocrystalline silicon substrate, in which the emitter and the collector areas, on opposite sides of the silicon substrate, were fabricated based on the ITO/n-Si junctions. The ITO film was deposited using the spray deposition technique described followed by a photolithographic formation of the emitter and the collector areas. The treatment in the H_2O_2 solution described above was applied to the silicon substrate. An ohmic n^+-contact (the base) was formed using local diffusion of phosphorous in the silicon substrate. The dependence of the collector current versus the collector-base voltage, using the emitter current as a parameter, are shown in

Figure 13, together with the emitter injection efficiency of the ITO/n-Si/ITO transistor (Malik et al., 2004).

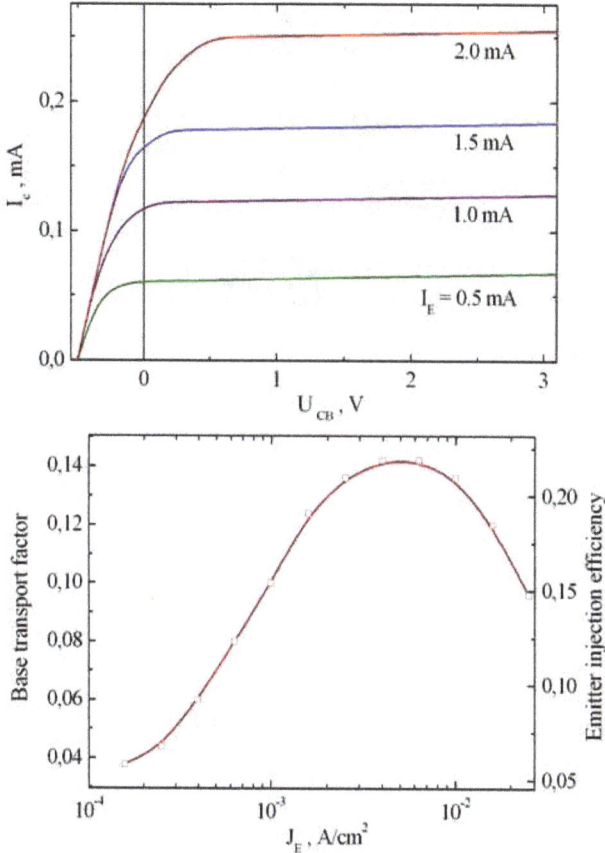

Fig. 13. Dependence of the collector current on the collector-base voltage (the emitter current is used as a parameter). The emitter injection efficiency of the ITO/n-Si/ITO transistor fabricated on a 10 Ω-cm silicon substrate is also shown. (Malik et al., 2004)

Hence, even in non-optimized transistors (wide base), an efficient hole injection of around 0.2 was observed. This is an obvious evidence of the existence of an inversion layer in the ITO/n-Si heterostructures with a barrier height of 0.9 eV. We can also present two indirect evidences of the p-n nature of the ITO/n-Si heterojunctions. The first one is based on the observation of an efficient radiation emission from the ITO/n-Si structures under a forward bias (Malik et al., 2004). In metal-semiconductor contacts operated as majority carriers' devices (described by the Schottky theory), the injection ratio does not exceed 10^{-4}. Thus, an efficient electroluminescence, in contrast to our devices, is not possible to observe. The next evidence is based on the observed modulation of the conductivity in the forward-biased ITO/n-Si diodes fabricated on high resistivity silicon, which operate as p-i-n diodes. So, the 0.9 eV barrier height belongs to an inversion ITO/n-Si heterojunction. This gives us the possibility to analyze theoretically such structures based on the well-known theory of p-n junctions.

4.3 Limit of applicability of the p-n model for the ITO/n-Si solar cells

Once we know the physical nature of the ITO/n-Si heterojunctions with extremely high potential barrier, it is possible to apply correctly the theory for their modelling, which is known as the theory of p-n based solar cells. The problem now is to find the range of resistivity of the silicon substrate on which the p-n theory can be applied to the ITO/n-Si heterojunction with extremely high potential barrier. Based on results published recently (Malik et al., 2008), the condition for strong inversion in the ITO/n-Si heterojunction requires that

$$\varphi_s \geq 2(E_F - E_i),\tag{1}$$

where

$$E_F - E_i = kT \ln(N_d / n_i),\tag{2}$$

φ_s is the surface potential at the Si/SiO$_x$ interface, k is the Boltzmann constant, T is the temperature, n_i is the intrinsic carrier concentration, and N$_d$ is the donor concentration in the n-Si substrate. On the other hand,

$$\varphi_s = \varphi_b - (E_C - E_F),\tag{3}$$

$$E_C - E_F = kT \ln(N_C / N_d),\tag{4}$$

where φ_b is the potential barrier for carriers from the ITO side of the structure, and N$_C$ is the effective density of states in the conduction band.

Moreover, the surface hole concentration is

$$p_s(x = 0) = (n_i^2 / N_d) \exp(\varphi_s / kT)\tag{5}$$

Combining equations (2)-(5), it is possible to obtain the surface concentration of the minority carriers at the Si/SiO$_x$ interface under strong inversion of the conductivity type:

$$p_s(x = 0) = (n_i^2 / N_C) \exp(\varphi_b / kT).\tag{6}$$

This concentration depends only on the barrier height and not on N$_d$. Figure 14 shows the two possible models in the space $p_s(x=0)/N_d$ vs. N$_d$ in the substrate for different barrier heights.

The two shaded areas are related to the two possible models: a Schottky model for $p_s(x=0)/N_d$ <0.01 and an induced p-n junction, in which $p_s(x=0)/N_d$ >10. For instance, at a barrier height of 0.7 V, the green line takes two intercepts: one with the border of the area that is related to the Schottky barrier model, and the other one with the border of the area that is valid for the p-n inversion model. Thus, for N$_d$>3x10^{14} cm^{-3} the structures behave as Schottky-barrier structures, whereas the structures with N$_d$<4x10^{13} cm^{-3}, behave as p-n

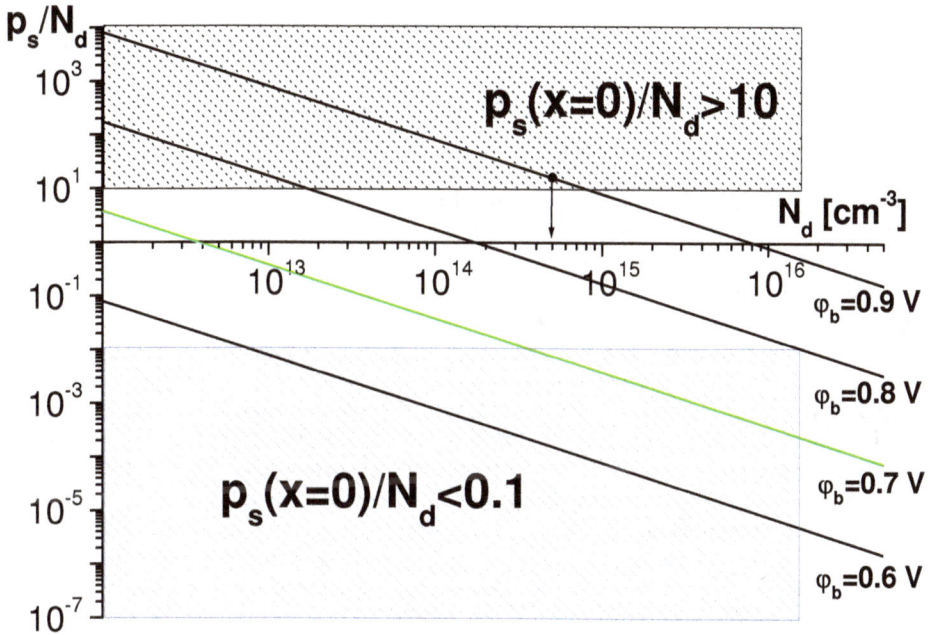

Fig. 14. Schematically representation of two possible models of the ITO/n-Si heterojunction in coordinates of $p_s(x=0)/N_d$ vs. concentration N_d in the silicon substrate. The different barrier height serves as a parameter (Malik et al., 2008)

junctions. With the potential barrier height of 0.9 eV achieved in this work, the structures may be considered as a symmetrical p-n ($p_s = N_d$) for N_d=8x10^{15} cm^{-3} (0.3 Ω-cm resistivity of the substrate), or as an asymmetrical p^+-n junctions ($p_s \geq 10N_d$) for N_d=8x10^{14} cm^{-3} (5 Ω-cm resistivity of the substrate). Due to the substrate resistivity used in this work, 10 Ω-cm (N_d=5x10^{14} cm^{-3}), our solar cells with a barrier height of 0.9 eV present an asymmetrical p^+-n junctions, and the theoretical analysis of such structures will be conducted based on the theory of p+-n junctions.

We underline again that the intermediate SiO$_x$ layer formed after the treatment of the silicon substrate in the H$_2$O$_2$ solution is sufficiently "transparent" for the carriers; then the tunneling current through this layer provides an ohmic contact between the ITO film and the surface-induced p^+-Si layer.

Thus, we can apply the diffusion theory of the p-n junction based solar cells for modelling the ITO/n-Si solar cells with a barrier height of 0.9 eV (the barrier height does not depend on the substrate carrier concentration) for a silicon substrate resistivity higher than 0.5 Ω-cm (or a carrier concentration lower than 8x10^{15} cm^{-3}).

4.4 ITO/n-Si solar cells: design, fabrication and characterization

The solar cells were fabricated using (100) n-type (phosphorous doped) single-crystalline silicon wafers with a 10 Ω-cm resistivity. Both sides of the wafer were polished. Standard wafer cleaning procedure was used. To form the barrier, an 80 nm-thick ITO film with a sheet resistance of 30 Ω/□ was deposited by spray pyrolysis on the silicon substrate treated in the H_2O_2 solution. This ITO thickness was chosen in order to obtain an effective antireflection action of the film. Metal, as an ohmic contact in the back side of the wafer, was deposited on an n^+-layer previously created by diffusion. The device area for measurements was 1-4 cm². Approximately 1 μm-thick Cr/Cu/Cr film was evaporated through a metal mask to create a grid pattern (approximately 10 grid-lines/cm). After fabrication, the capacity-voltage characterization was conducted to control the value of the potential barrier. Then the following parameters were measured under AMO and AM1.5 illumination: open circuit voltage V_{oc}, short circuit current I_{sc}, fill factor FF, and efficiency. No attempt was made to optimize the efficiency of the cells by improving the collection grid. The series resistance (R_s) of the cell was measured using the $R_s=(V-V_{oc})/I_{sc}$ relationship (Rajkanan, Shewchun, 1979), where V is the voltage from the dark I-U characteristics evaluated at $I=I_{sc}$.

It was shown above that the ITO/n-Si heterostructures with a potential barrier height at the silicon surface of 0.9 eV behave as pseudo classical diffusion p-n junctions. Thus, it is expected that the diffusion of holes in the silicon bulk dominates the carrier transport instead of the dominance of the thermo-ionic emission in the Schottky and the metal/tunnel oxide/semiconductor structures. A straightforward measurement of the dependence of the dark current on temperature is, in principle, sufficient to identify a bipolar device in which the thermo-ionic current is negligible in comparison to the minority-carrier diffusion current J_d (in units of current density). A simple Shockley's analysis of the p-n diode including the temperature dependence of the silicon parameters (diffusion length, diffusion coefficient, minority carrier life-time, and the intrinsic concentration) (Tarr, Pulfrey, 1979) shows that

$$J_d = J_{0d}[\exp\ (qV/kT)-1] \tag{7}$$

and

$$J_{0d} \propto T^{\gamma} \exp\ (-E_{g0}/kT), \tag{8}$$

where $\gamma = 2.4$ and $E_{g0} = 1.20$ eV.

From Eq.(8) it can be seen that the plot $\log(J_{0d}/T^{\gamma})$ vs. $1/T$ should produce a straight line, and that the slope of this line should be the energy E_{g0}. In the case of MS and MIS devices this slop must be equal to the value of the barrier φ_b.

Usually, the series resistance of the device affects the I-V characteristics at high forward current densities. To prevent this effect, we must measure the J_{sc} vs V_{oc} dependences (Rajkanan, Shewchun, 1979). The photogenerated current is equal to the saturation photocurrent. For minority-carrier MIS diode with a thin insulating layer (Tarr, Pulfrey, 1979)

$$J_{sc} = J_{rg}(V_{oc}) + J_d(V_{oc}). \tag{9}$$

For an increasing bias, J_d increases faster than the recombination current density J_{rg}; in the high illumination limit we should have

$$J_{sc} = J_{0d} \exp(qV_{oc}/nkT), \tag{10}$$

which gives an n factor approximately equal to 1.

Figure 15 shows the measured dependence of J_{sc} on V_{oc} at room temperature. The value of J_{0d} in (10) was determined by measuring J_{sc} and V_{oc} at different temperatures, and under illumination with a tungsten lamp. An optical filter was used to prevent the heating of the cell by the infra-red radiation. For each J_{sc} - V_{oc} pair lying in the range where $n \approx 1$, $J_{0d}=J_{02}$ was calculated from (10). After making the correction for the T^{γ} factor appearing in Eq.(8), the J_{0d} values were plotted as a function of the reciprocal temperature, as shown in the insert of Figure 15.

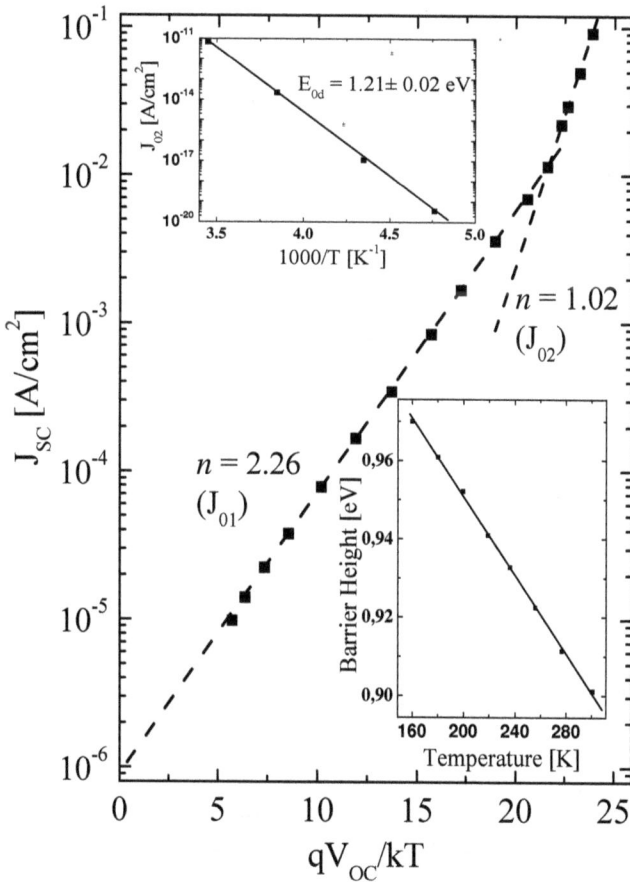

Fig. 15. Measured dependence of J_{sc} on V_{oc} at room temperature, and calculated dependence of the current density $J_{02}=J_{0d}$ at high illumination level corrected for the T^{γ} factor, as function of reciprocal temperature for ITO/n-Si solar cells with the barrier height of 0.9 eV. The dependence of the barrier height on temperature is also shown in the insert

The slope of the J_{02} vs. $1/T$ line was found to correspond to an energy E_{g0} from Eq.(8). It can be concluded that for high current densities the current in the cell is carried almost exclusively by holes injected from the ITO contact that later diffuse into the base of the cell. The output characteristics of the ITO/n-Si solar cell measured under AM0 and AM1.5 illumination conditions, as well as the calculated dependence of output power of the cell versus the photocurrent, are shown in Figure 16.

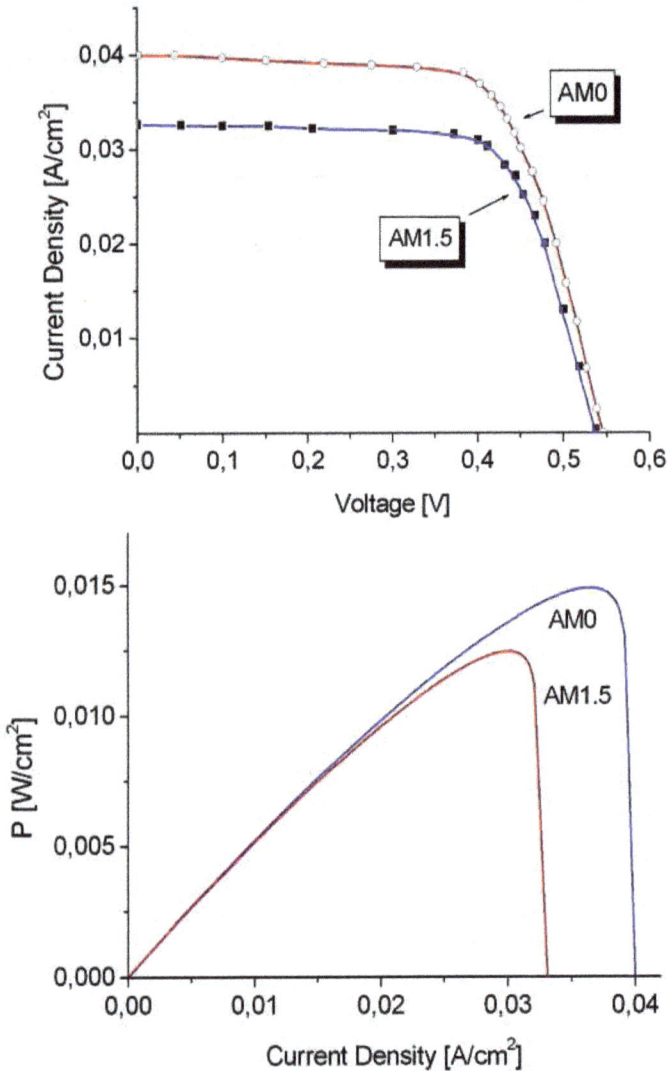

Fig. 16. *I-V* characteristics (above graph) of the ITO/n-Si solar cell measured under AM0 and AM1.5 illumination conditions, and the calculated dependence of the output power of the cell (below graph) versus the photocurrent

The fill factor (FF) and the efficiency calculated from these characteristics are 0.68 and 10.8% for AM0 illumination conditions; and 0.68 and 12.1% for AM1.5 illumination conditions. The fill factor and efficiency obtained are not optimized because of the cell design and the used silicon substrate with relatively high resistivity. Below, a theoretical analysis followed for increasing the output parameters of the cells using silicon substrates with lower resistivity is presented.

4.5 Optimization of the output characteristics of the cells: theoretical analysis

Recently, a detailed theoretical modelling of the ITO/n-Si solar cells has been reported (Malik et al, 2008). Based on these published results, here we show the most important conclusions; further details can be consulted in that work.

For all the calculations, the thickness of the silicon substrate, sheet resistance and thickness of the ITO film were taken as d=500 μm, 30 Ω/□, and t=80 nm, respectively. We considered the case when the diffusion length of minority carriers is shorter than the thickness of the silicon substrate, and assumed that the carrier recombination rate at the back contact of the silicon substrate is infinite. The total series resistance of the cell with an area of 1 cm² used for these calculations was taken as 1.8 Ω.

In order to calculate the theoretical parameters of the solar cells we assumed also that the equation for the I-V characteristic for an illuminated cell (Sze & Ng, 2007) is

$$\ln\left(\frac{J + J_{sc}}{J_0} - \frac{V - JR_s}{J_0 R_{sh}} + 1\right) = \frac{q}{\gamma kT}(V - JR_s), \tag{11}$$

where J is the current density, J_0 the saturation dark current density, J_{sc} is the short-circuit current density, V is the output voltage, R_s and R_{sh} are the series and shunt resistances, and γ is the "ideality" factor of the solar cell. According to our experimental results, γ was taken as 1 for the calculations.

In order to calculate the photocurrent density we integrated the next equation based on the spectral distribution of the incident solar radiation, and the parameters of silicon (absorption coefficient $\alpha(\lambda)$, diffusion length for minority carriers L_p, and thickness of the silicon substrate d):

$$J_{sc} = q \int_{\lambda 1}^{\lambda 2} \left\{ (1-R)_\lambda F_\lambda \left(\frac{\alpha L_p}{\alpha^2 L_p^2 - 1} e^{-\alpha W}\right) \times \left(\alpha L_p - \frac{\cosh\left(\frac{d}{L_p}\right) - e^{-\alpha d}}{\sinh\left(\frac{d}{L_p}\right)} \right) + (1-R)_\lambda F_\lambda (1 - e^{-\alpha W}) \right\} d\lambda \tag{12}$$

where q is the electron charge, W is the depletion width in the silicon substrate, and $R(\lambda)$ is the spectral reflectance from the ITO/Si interface calculated from the optical constants of silicon and ITO (Malik et al, 2008).

The spectral distribution F_λ of the solar radiation, which are related to the AM0 (136 mW/cm²) and AM1.5 (100 mW/cm²) conditions, have been used in the calculations according to the 2000 ASTME-490-00 and ASTM G-173-03 standards, respectively.

The values of the open-circuit voltage under AM0 and AM1 conditions were calculated according to the equation

$$V_{oc} = \frac{\gamma kT}{q} \ln\left(\frac{J_{sc}}{J_0} + 1\right), \tag{13}$$

where the saturation dark current density J_0 is calculated from the equation

$$J_0 = qn_i^2 \frac{D_p}{N_d L_p} coth \frac{d}{L_p} \tag{14}$$

Here, n_i and N_d are the intrinsic and donor concentrations in the silicon substrate, respectively, and D_p is the diffusion coefficient for holes.

Figure 17 shows the experimental (dots) and calculated (lines) I-V characteristics of the solar cell (using equation (11)) with an area of 1 cm² fabricated on 10 Ω-cm silicon under both AM0 and AM1.5 illumination conditions. Initially, these characteristics were calculated using J_{sc} = 40 mA/cm², R_s = 1.8 Ω, and R_{sh} = ∞. Then, in order to improve the fitting with the experimental results, the calculated characteristics were corrected using R_{sh}= 300 Ω. One can see an excellent coincidence between the experimental and calculated characteristics, as well as for the parameters of the cell (fill factor F.F. and conversion efficiency η).

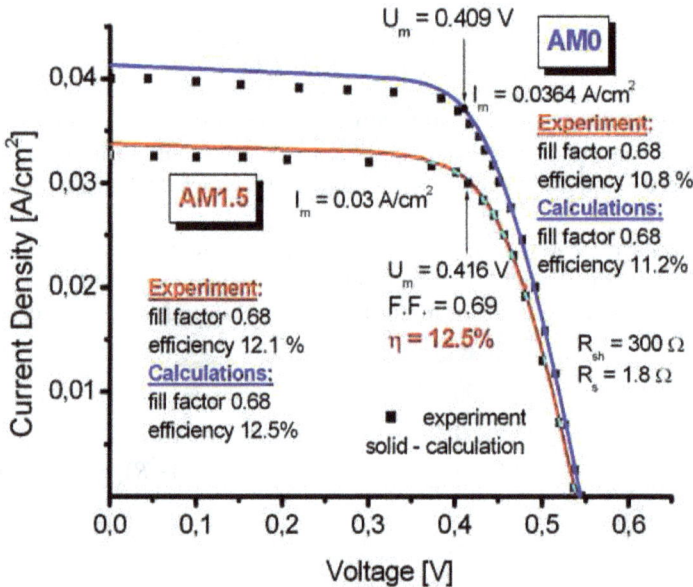

Fig. 17. Experimental (dots) and calculated (solid lines, using equation (11)) I-V characteristics of the solar cell with an area of 1 cm², and fabricated on a 10 Ω-cm silicon substrate, under both AM0 and AM1.5 illumination conditions

Figure 18 (above graph) shows the dependence of the short-circuit current density J_{sc} on the diffusion length in the silicon substrate, under AM0 and AM1.5 illumination conditions.

The values of the open-circuit voltage under the same conditions were calculated according to equations. (13) and (14). From equation (14), the value of J_0 decreases with the resistivity ρ of the silicon substrate. The calculated dependence for the open-circuit voltage (V_{OC}) on the resistivity of the Si substrate is also shown in Figure 18 (below graph).

The calculations show that the conversion efficiency of the ITO-SiO$_x$-nSi solar cells can be improved by using silicon with a lower resistivity. Under the AM1.5 conditions, the calculated dependences of the open circuit voltage, fill factor, and efficiency on the resistivity of the silicon substrate, are shown in Figure 19. The surface recombination velocity (S_p) was taken into account for these calculations. The value of S_p for the ITO/n-Si heterostructures under investigation, and determined from the analysis of the transistor structures, was 500 cm/s approximately.

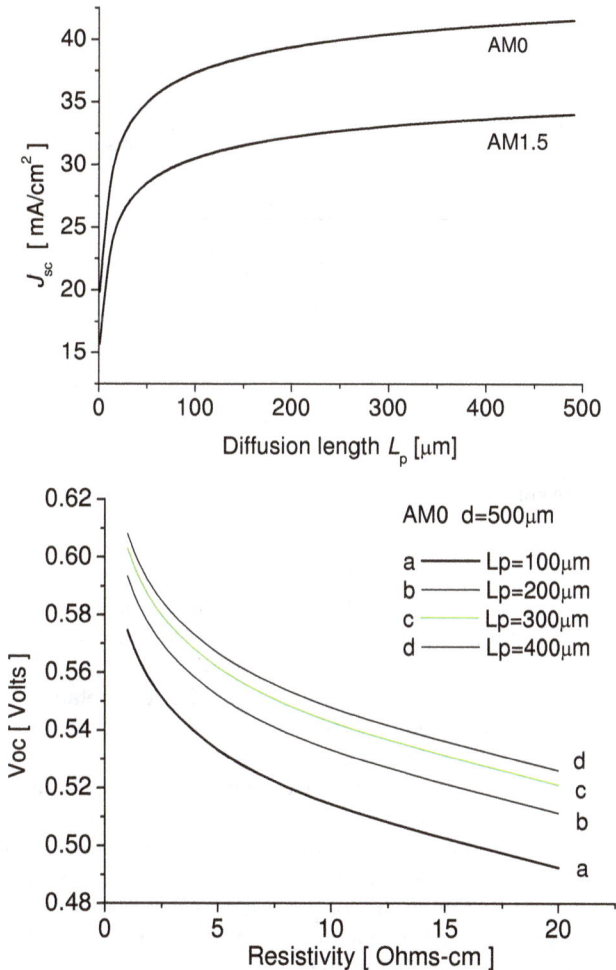

Fig. 18. Calculated dependence of J_{sc} and V_{OC} for the ITO/n-Si solar cells

Solar cells fabricated on silicon substrates with a resistivity of 1Ω-cm and a hole diffusion length of L_p= 200 μm may present an efficiency of 14 %. For an experimentally found potential barrier of 0.9 eV it is not possible to achieve a further reduction of the silicon resistivity for structures with a *p-n* inversion layer or minority carrier devices. Such structures are majority carrier devices, and their properties are described by the theory of Schottky barriers. In such cases, a lower efficiency is expected due to a higher saturation current. Solar cells using FTO films present similar characteristics.

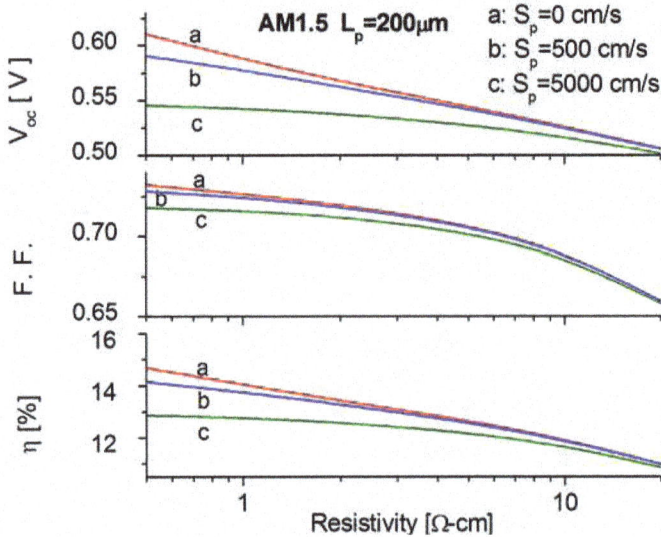

Fig. 19. Calculated dependences of the series resistance, fill factor, and efficiency of the cells, on the resistivity of the silicon substrate.

5. Conclusions

ITO-nSi solar cells have been produced using a spraying technique. Transparent and conductive tin-doped indium oxide films, as well as fluorine-doped tin oxide films, presenting excellent structural, optical and electrical parameters, were fabricated using a very simple, low cost, and no-time consuming method. The cells obtained in such a way can be considered as structures presenting an inversion p-n junction. Under the AM0 and AM1.5 solar illumination conditions, the efficiency is 10.8% and 12.2%, respectively. The theoretical modelling based on p-n solar cells show an excellent coincidence between the theoretical and the experimental results. It is also shown that using 1 Ω-cm silicon substrates is a promising alternative for obtaining solar cells with 14% efficiency under AM1.5 illumination conditions. The use of substrates with a lower resistivity leads to a reduction of the conversion efficiency due to the formation of Schottky barriers, which gives place to a higher saturation dark current than that presented by p-n structures. The fabrication of reported solar cells is more controllable than that needed for obtaining metal-insulator-silicon solar cells because of the necessity of controlling a very thin (nearly 2 nm) intermediate oxide layer on the silicon substrate. Moreover, a detailed theoretical analysis (Shewchun et al., 1980) shows a higher efficiency for p-n inversion solar cells in comparison with those based on majority-carrier MIS structures.

6. References

Ashok, S.; Sharma, P. & Fonash, S. (1980). Spray-deposited ITO-silicon SIS heterojunction solar cells. *IEEE Trans. Electron. Dev.*, Vol.ED-27, N.4, 725-730, ISSN 0018-9383

Dawar, A. & Joshi J. (1984). Semiconducting transparent thin films: their properties and applications. *J. Mater. Sci.*, Vol.19, 1-23, ISSN 0022-2461

DuBow, J.; Burk, D. & Sites, J. (1976). Efficient photovoltaic heterojunctions of indium tin oxides on silicon. *Appl. Phys. Lett.*, Vol.29, N.8, 494-496, ISSN 0003-6951

Feng, T.; Ghosh, A. & Fishman, G. (1979). Efficient electron-beam-deposited ITO/n-Si solar cells, *J. Appl. Phys.*, Vol.50, N.7, 4972-4974, ISSN 0022-3727

Fukano, T.; Motohiro, T. & Ida, T. (2005). Ionization potential of transparent conductive indium oxide films covered with a single layer of fluorine-doped tin oxide nanoparticles grown by spray pyrolysis deposition, *J. Appl. Phys.*, Vol.97, N.8, Nanoscale science and design, ISSN 0022-3727

Fujiwara, N.; Fujinaga, T.; Niinobe, D.; Maida, O.; Takahashi, M. & Kobayashi, H. (2003). Passivation of defect states in Si and Si/SiO$_2$ interface states by cyanide treatment: improvement of characteristics of pin-junction amorphous Si and crystalline Si-based metal-oxide-semiconductor junction solar cells. *Acta Phys. Slovaca*, Vol.53, N.3, 195-205, ISSN 0323-0465

Granqvist, C. (1993). Transparent Conductive Electrodes for Electrochromic Devices: A Review. *Appl. Phys.*, Vol.A57, 19-24, ISSN 0947-8396

Gouskov, L.; Saurel, J.; Gril, C.; Boustani, M. & Oemry, A. (1983). Sprayed indium tin oxide layers: Optical parameters in the near-IR and evaluation of performance as a transparent antireflecting and conducting coating on GaSb or Ga$_{1-x}$Al$_x$Sb for IR photodetection. *Thin Solid Films*, Vol.99, N.4, 365-369, ISSN 0040-6090

Haacke, J. (1976). New figure of merit for transparent conductors. *J. Appl. Phys.*, Vol.47, 4086-4089, ISSN 0022-3727

Hamberg, J. & Granqvist, C. (1986). Evaporated Sn-doped In2O3 films: basic optical properties and applications to energy-efficient windows. *J. Appl. Phys.*, V.60, n.11, R13, ISSN 0022-3727

Hartnagel, H.; Dawar, A.; Jain, A. & Jagadish, C. (1995). *Semiconducting Transparent Thin Films*, IOP Publishing Ltd., ISBN 0 7503 03220, Bristol UK

Malik, A.; Baranyuk, V. & Manasson, V. (1979). Solar cells based on the SnO$_2$-SiO$_2$-Si heterojunction. *Appl. Sol. Energy*, N. 2, 83-84, ISSN 0003-701X

Malik, A.; Baranyuk, V. & Manasson, V. (1980). Improved model of solar cells based on the In2O3/ SnO$_2$-SiO$_x$-nSi structure. *Appl. Sol. Energy*, N.1, 1-2, ISSN 0003-701X

Malik, O.; Grimalsky, V.; Torres-J., A. & De la Hidalga-W, J. Room Temperature Electroluminescence from Metal Oxide-Silicon. *Proceedings of the 16th International Conference on Microelectronics (ICM 2004)*, pp. 471-474, ISBN 0-7803-8656-6, Tunis, December 06-08, 2004, IEEE, Tunisia

Malik, O.; Grimalsky, V. & De la Hidalga-W, J. (2006). Spray deposited heavy doped indium oxide films as an efficient hole supplier in silicon light-emitting diodes. *J. Non-Cryst. Sol.*, Vol.352, 1461-1465, ISSN 0022-3093

Malik, O.; De la Hidalga-W, J.; Zúñiga-I, C. & Ruiz-T, G. (2008). Efficient ITO-Si solar cells and power modules fabricated with a low temperature technology: results and perspectives. *J. Non-Cryst. Sol.*, Vol.354, 2472-2477, ISSN 0022-3093

Manifacier, J. & Szepessy, L. (1977). Efficient sprayed In2O3:Sn n-type silicon heterojunction solar cell. *Appl. Phys. Lett.*, Vol.31, N.7, 459-462, ISSN 0003-6951

Manifacier, J.; Fillard, J. & Bind J. (1981). Deposition of In2O3-SnO2 layers on glass substrates using a spraying method. *Thin Solid Films*, Vol. 77, N.1-3, 67-80, ISSN 0040-6090

Moholkar, A.; Pawar, S.; Rajpure, K.; & Bhosale, C. (2007). Effect of solvent ratio on the properties of highly oriented sprayed fluorine-doped tin oxide thin films. *Mater. Lett.*, Vol.61, N.14-15, 3030-3036, ISSN 0167-577X

Nagatomo, T.; Endo, M. & Omoto, O. (1979). Fabrication and characterization of SnO2/n-Si solar cells, *Jpn. J. Appl. Phys.*, Vol.18, 1103-1109, ISSN 0021-4922

Nagatomo, T.; Inagaki, Y.; Amano, Y. & Omoto, O. (1982). A comparison of spray deposited ITO/n-Si and SnO2/n-Si solar cells, *Jpn. J. Appl. Phys.*, Vol.21, N. 21-2, 121-124, ISSN 0021-4922

Nakasa, A.; Adachi, M.; Suzuki, E.; Usami, H. & Fujimatsu, H. (2005). Increase in conductivity and work function of pyrosol indium tin oxide by infrared irradiation, *Thin Solid Films*, Vol.84, N.1-2, 272-277, ISSN 0040-6090

Nath, P. & Bunshah, R. (1980). Preparation of In_2O_3 and tin-doped In_2O_3 films by novel activated reactive evaporation technique. *Thin Solid Films*, Vol.69, N.1, 63-68, ISSN 0040-6090

Rajkanan, K. & Shewchun, J.(1979) A better approach to the evaluation of the series resistance of solar cells, *Sol. St. Electron.*, Vol.22, N.2-E, 193-197, ISSN 0038-1101

Saxena, A.; Singh, S.; Thangaraj, R. & Agnihotri O. (1984). Thickness dependence of the electrical and structural properties of In_2O_3:Sn films, *Thin Solid Films*, Vol.117, N.2, 95-100, ISSN 0040-6090

Shewchun, J.; Burc, D. & Spitzer, M. (1980). MIS and SIS solar cells, *IEEE Trans. Electron. Dev.*, Vol.ED-27, N.4, 705-716, ISSN 0018-9383

Song, G.; Ali, M. & Tao, M. (2008). A high Schottky barrier between Ni and S-passivated n-type Si (100) surface. *Sol. St. Electron.*, Vol.52, 1778-1781, ISSN 0038-1101

Sze, S. & Ng, K. (2007). *Physics of semiconductor devices*, 3rd ed., John Wiley and Sons, ISBN 9780471143239, N.Y.

Tarr, N. & Pulfrey, D. (1979). New experimental evidence for minority-carrier MIS diodes. *Appl. Phys. Lett.*, V.34, N.4, 15 February 1979, 295-297, ISSN 0003-6951

Theuwissen, A. & Declerck, G. (1984). Optical and electrical properties of reactively d. c. magnetron-sputtered In_2O_3:Sn films. *Thin Solid Films*, Vol.121, N.2, 109-119, ISSN 0040-6090

Trmop, R.; Hamers, R. & Demuth, J. (2005). Si (100) dimmer structure observed with scanning tunneling microscopy, *Phys. Rev. Lett.*, Vol.55, N.12, 1303- 1308, ISSN 0031-9007

Vasu, V. & Snbrahmanyam, A. (1990). Reaction kinetics of the formation of indium tin oxide films grown by spray pyrolysis. *Thin Solid Films*, Vol.193-194, n.2, 696-703, ISSN 0040-6090

Verhaverbeke, S.; Parker, J. & McConnell, C. (1997). The role of HO_2^-: in SC-1 cleaning solutions. In: Mat. Res. Soc. Symp. Proc., Vol. 477: Science and Technology of Semiconductor Surface Preparation, 47-56, MRS, ISBN 1-55899-381-9, N.Y.

6

Charge Carrier Recombination in Bulk Heterojunction Organic Solar Cells

Gytis Juška and Kęstutis Arlauskas
Vilnius University
Lithuania

1. Introduction

Photovoltaic phenomenon was first observed by E. Becquerel (Becquerel) in 1839. He observed the electric current-lit silver electrode, immersed in the electrolyte. In 1894, taking advantage of the observed photoconductivity phenomenon in amorphous selenium the semiconductor solar cell was developed.

The very first silicon p-n junction solar cell was made in 1954, energy conversion efficiency of which was 6% and the energy price $200/W did not seem promising for wide application. Later, the development of satellites needed to provide sustainable energy sources and the cadmium sulfid, cadmium telluride, gallium arsenide and more efficient solar cells of other materials were created.

The first solar cell breakthrough was something like of the 1970 year, feeling the lack of oil, which oncreased interest in alternative energy sources. The basic raw materials, in addition to crystalline silicon, a polycrystalline silicon, were also amorphous silicon and other, suitable for thin solar cells, materials. Although, due to the high cost of these energy sources, extracted energy was only a small part of total energy production, but the lending spread as energy sources in various areas of small devices such as mobile phone, calculators, meteorogical instruments, watches and so on. A solar powered cars and even solar powered aircraft were constructed. Major Solar cells used for the purification of salt water, as well as supply power to isolated objects: mountains, islands or jungle living population.

The second and much greater solar energy use breakthrough occurred in the first decade of the twenty-first century. This is caused by the earth's climate warming due to the increasing threat of thermal energy and the increasing CO_2 in the atmosphere. Many governments in many ways stimulated the solar energy lending. Germany in the decade from 1994 to 2004, installed as much as 70 times more solar energy equipment, and now is installed more than 1GW: produced over 3TWh energy, which cost around 0.5 €/kWh. In Japan solar power energy is less costly than the heat. The main price of solar energy is caused by the installation consts - ~ 1€/W. Till 2004 there have already been installed over 1GW, while in 2006, the world's installed 6.5 GW. In 2007, the European Union in the fight against climate warming threat committed by 2030 to achieve that 25% of the total energy from alternative sources, mainly from the Sun. It should be around 1200 GW, the cost should not exceed 0.1 €/kWh. Another reason for the needed alternative energy sources is projected oil and gas resource depletion.

Crystalline silicon still remains the unrivaled leader in the development of solar cells. However, the demant of renewable energy sources stimulated a search for a new, low-cost technologies and materials. Hydrogenated amorphous silicon (a-Si:H) has long been regarded as one of the most promising materials for development of cheap, lightweight and technologicall solar cells. However, a Si:H solar cells degraded in high intensity-light. Thus, forward-looking, more efficient microcrystalline (μc-Si:H) and nanocrystalline silicon (nc-Si:H) solar cells began to compete successfully with a-Si:H.

The first organic materials were investigated for more than a hundred years ago and for a long time the widest application, in scope of optoelectronics, was electrography. However, in 1977 A. J. Heeger, A. G. MacDiarmid and H. Shirakawa showed that the π-conjugated polymers can be doped, and change the properties of substances. This work demonstrated the possibility use polymers to create optoelectrical devices, resulted in huge interest and in 2000 was awarded the Nobel Prize. During the period from 1977 on the base of π-conjugated polymers has been built a number of electronic and optoelectronic devices: diodes, field effect transistors, sensors, photodiodes, etc. On 1993 - 2003 years π-conjugated polymers have been investigated in order to create a light-emitting diodes (OLED) and their systems, and these studies culminated in the creation of a colour OLED matrix, which is adapted to different types of displays. Recently, organic polymers mainly involved studies of organic solar cells and other organic electronics appliances, effectiveness of which is determined by the drift and recombination of charge carriers.

In order to develop efficient solar cells it is necessary the maximum possible the light absorption, the carrier photogeneration quantum efficiency, and that all photogenerated carriers be collected in a solar cell electrodes. The collection of charge carriers depends on their mobility and recombination. Thus, the investigations of carrier mobility and their density dependencies on the electric field, temperature and material structure are essential for the formation of understanding of charge carrier transport in these materials, which is essential to find effective new inorganic and organic materials and to development of new optoelectronic structures.

One of the main factors limiting efficiency of organic solar cells (OSC) is charge carrier recombination. In crystals, where the carrier location uncertain, recombination is caused by the probability to transfere energy: or emit photon - radiation recombination, or to another electron - Auger recombination, or induction phonons through the deep states. The latter depends on the density of deep states. In disordered structures, with a lot of localized states, should be very rapid recombination, but there recombinationis caused by the meeting probability of electron and hole in space, as the only their meeting at a distance closer than the Coulomb radius causes their recombination (named Langevin), likely as gemini recombination. It is valid only if the energy dissipation or jump distance is less than the Coulomb radius. Thus, the Langevin bimolecular recombination is ordained by the mutual Coulomb attraction drift time, because under this attraction electron is moving toward the nearest hole, while at the same time, due to diffusion, with equal probability in any direction. The Langevin recombination time can be expressed as:

$$\tau_L = \int_0^r \frac{dx}{(\mu_n + \mu_p)F} = \frac{\varepsilon\varepsilon_0}{e(\mu_n + \mu_p)n} \tag{1}$$

Here μ_n, μ_p are electron and hole mobility, respectively; $F = e/4\pi\varepsilon\varepsilon_0 x^2$ is strength of Coulomb electric field; n is density of charge carriers ($1/n = 4\pi r^3/3$), and r is a mean distance between electrone and hole. Thus, from the expression of Langevin bimolecular recombination coefficient $B_L = e(\mu_n + \mu_p) / \varepsilon\varepsilon_0$ it is clearly seen that recombination is caused by the features of charge carrier transport. In bulk heterojunction organic solar cells the reduced Langevin recombination is observed.

In this work we describe methods of investigation of charge carrier recombination in disordered structures, where stochastic transport of charge carriers complicates interpretation of experimental results: integral time of flight (i-TOF) (Juška et, 1995), using of which allows easily estimate the temperature dependence of recombination coefficient; charge carriers extraction by linearly increasing voltage (CELIV) (Juška et, 2000, a), which allows independently measure relaxation of density and mobility of photoexcited charge carriers; double injection current transient (DoI) (Juška et, 2005; Juška et, 2007), which is additional method of investigation of charge carrier recombination and, which allows to measure dependence of recombination coefficient on electric field.

In this study we represent how using current transient methods may be cleared up the features of charge carrier transport and recombination in disordered inorganic and organic materials. The microcrystalline silicon and π-conjugated polymers have been investigated as a typical inorganic and organic material.

2. Investigation methods

The disordered structure of material causes that mobility of charge carriers is low, because their motion is slowed down by the interaction with spectrum of the local states. Thus, the classical investigation methods: the Hall and magnetoresistance measurements are invalid. The carrier transport in disordered inorganic and organic materials, conductivity (σ) of which is low, is studied using time-of-flight (TOF) method. However, the conductivity of many π-conjugated polymers is high and does not fulfill the latter condition. Thus, for their investigation has been adapted and refined microcrystalline hydrogenated silicon (μc-Si:H) used the extraction of charge carriers by linearly increasing voltage (CELIV) method. The latter method allows to investigate the transport properties of charge carriers both in conductive and low conductivity materials. For investigation of charge carrier transport and recombination the double injection current (DoI) transient method is promising as well.

2.1 TOF method

Time-of-flight method is widely used for investigation of transport, trapping-retrapping and recombination of charge carriers in disordered materials and structures. This method is applicable only for investigation of low conductivity materials, i.e. where the Maxwell relaxation time exceeds the duration of transit time (t_{tr}) of charge carriers through the interelectrode distance (d):

$$\tau_\sigma = \frac{\varepsilon\varepsilon_0}{\sigma} \gg t_{tr} = \frac{d^2}{\mu U} \tag{2}$$

TOF method is based on the current transient measurement when photogenerated of the same sign charge carriers is moving in the electric field (E) created in the interelctrode distance (d) of the sample and during a drift time (t_{tr}) the package achieves an opposite electrode. The simplicity and efficiency of method meant that it is a widely used for study of mobility (μ), trapping (τ_t) and lifetime of charge carriers (τ) in low conductivity ($\tau_\sigma > t_{tr}$)

materials. Low conductivity of material ensures that during the drift of photogenerated charge carriers through interelectrod distance the density equilibrium charge carriers will be too low to redistribute the electric field inside the sample, and the electric field will be steady at the moment of charge carrier photogeneration, i.e. $RC < t_{del} < \sqrt{\tau_\sigma t_{tr}}$ (here R is total resistance of measurement system and sample elctrodes, C is geometric capacitance of sample). TOF method, dependently on amount of initial injected charge (Q_0) and, also, on characteristic time RC of measurement system, is devided into a number of regimes.

Small charge drift currents (SCDC). This regime is ensured when an amount of photogenerated charge is much less than an amount of charge on sample electrodes at given voltage (U_0), i.e. $eL = Q_0 << CU_0$. Here L amount of charge carriers photogenerated by pulse of light. In this regime there are a few cases:

a. *current (diferencial) regime ($t_{tr} > RC$).* In case of strong absorption of light ($\alpha d >> 1$, α is absorption coefficient) and nondispersive transport, the shape of pulse of photocurrent transient is close to rectangular, duration of which is t_{tr} (Fig. 1a, $L = 0,3$), and form the area of current transient the Q_0 can be estimated. In case of weak absorption of light ($\alpha d << 1$), charge carriers are photogenerated in the bulk of sample, thus, the shape of photocurrent pulse is triangular, which's duration is t_{tr}, and area is equal $Q_0/2$. The dispersive transport of charge carriers, due to dependence of charge carrier mobility on time, causes that pulse of current transient did not demonstrate obvious break points, form which will be possible to estimate t_{tr} (even if $\alpha d >> 1$). In this case, if the current transient is represented by a double-log scale ($lgj = f(lgt)$), the turning point corresponds the t_{tr}.

b. *charge (integral) regime ($t_{tr} < RC$).* Even in case of strong absorption of light and nondispersive transport of charge carriers the shape of photocurrent pulse is not so informative as in current regime (Fig. 1a, $L = 0,3$): the drift time of charge carrier package is estimated as halftime ($t_{1/2}$) of rise time of photocurrent pulse, i.e. $t_{tr} = 2\,t_{1/2}$. The magnitude of photocurrent pulse is equal to amount of charge (Q) collected onto the sample electrodes during the charge carrier drift time. In case of bulk absorption of light, the magnitude of photocurrent pulse is equal $Q/2$, and $t_{tr} = 3,41\,t_{1/2}$

If the voltage of backward direction is applied onto solar cell electrodes and, by short pulse of light the charge carrier pairs are photogenerated, the photocurrent pulse of their drift is observed, from which's duration (t_{tr}) the mobility of the charge carriers of the same polarity as illuminated electrode is estimated. An amount of drifting charge carriers is estimated from the area of photocurrent pulse, from which, when amount of absorbed quanta of light is known, the quantum efficiency is evaluated (Fig.1).

In case of trapping with characterstic trapping time τ or in case of stochastic transport, after photogeneration, the shape of photocurrent pulse is decreasing, and, from the area of photocurrent pulse, estimated dependence of amount of photogenerated charge carriers on voltage follows Hecht's dependence (Eg. (3)). From the latter dependence the $\mu\tau$-product, which determines both the diffusion and drift lengths of charge carries, and causes effectiveness of solar cell, is estimated.

$$\frac{N}{N_0} = \frac{\mu\tau E}{d}\left(1 - \exp\left(-\frac{d}{\mu\tau E}\right)\right) \tag{3}$$

Space charge limited photocurrent (SCLP). In this case an amount of phtogenerated charge is higher than charge on sample electrodes at U_0, i.e. $Q_0 \gg CU_0$. The shape of photocurrent pulse depends on Q_0 (Fig. 1a), and strongly absorbed light ($\alpha d \gg 1$) creates reservoir of charge carriers at the illuminated elektrode, from which not more than CU_0 charge package can drift to the opposite elektrode. This package is moving in growing electric field, thus, in case of nondispersive transport and when $t_{tr} > RC$, drift time is $t_{SCLC} = 0{,}78\, t_{tr}$, which is estimated from the spike of current transient (Fig. 1a). When $t > t_{SCLC}$, current flows until the whole charge is extracted from reservoir and the second turning point, at extraction time (t_e), appears on the pulse of photocurrent. An amount of charge extracted from the reservoir (Q_e), as well as t_e, depend on recombination speed of charge carriers in reservoir.

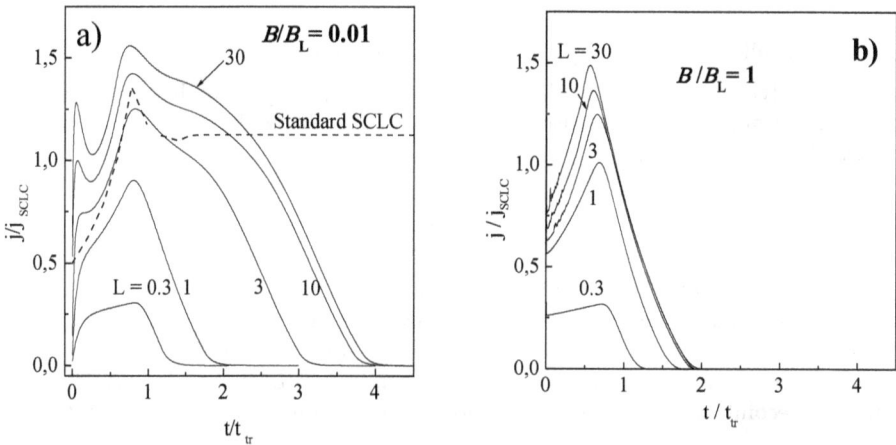

Fig. 1. Numerically modelled photocurrent transients of charge carrier drift dependence on exciting light intensity in case when $B/B_L = 0.01$ (a), and when Langevin recombination prevails (b). Density of photogenerated charge carriers is normalised to amount of charge on sample electrodes in SCLC regime

For investigation of charge carrier recombination by photocurrent transient methods the dependence of collected onto sample electrodes charge on intensity of photoexciting light pulse is measured (Pivrikas et, 2005). When, due to increasing intensity of light pulse, the amount of photogenerated charge achieves amount of charge carriers on sample electrodes ($Q_0 = CU_0$), the TOF regime changes from small charge drift current (SCDC) to space charge limited current (SCLC) (Fig. 2a). Further increase of light pulse intensity not follows by increase of photocurrent, but increases the duration ($t_e \geq t_{tr}$) of photocurrent pulse, which is caused by the extraction of charge carriers from reservoir. The faster charge carrier recombination in reservoir, the shorter extraction time (t_e), and, when recombination is very fast, $t_e \rightarrow t_{tr}$. Thus, the dependence of t_e on intensity of exciting light pulse L gives information about recombination process in charge carrier resrvoir: dependence as $t_e(L) \approx \ln L$ indicates that monomolecular recombination prevails; if, at high intensity of light pulse, t_e saturates with L, than the bimolecular or of higher order of charge carrier recombination prevails.

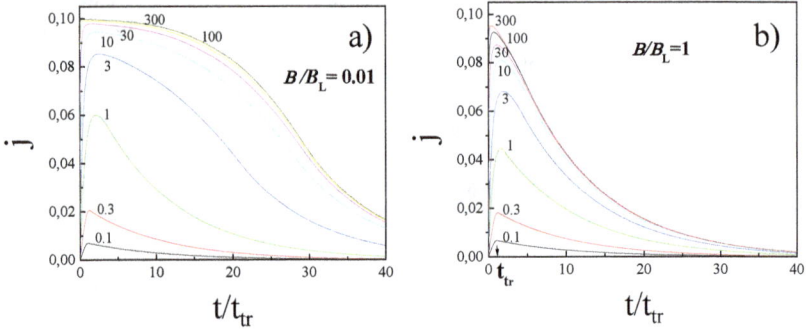

Fig. 2. Numerically modelled integral TOF current transients ($RC = 10\ t_{tr}$)

In organic polymers the bimolecular recombination typically is of Langevine-type. The photocurrent transients of this case are shown in Fig. 2b, and the maximal amount of extracted charge is estimated as:

$$\frac{Q}{CU} = 1 - \exp\left(-\frac{eL}{CU}\right) \qquad (4)$$

The maximal amount of extracted charge $Q = CU$.

When the bimolecular recombination is weaker than Langevin's one, from the saturation of extraction time, which is estimated as difference of photocurrent pulse halwidths at space charge limited and at small charge regimes, i.e. $t_e = t_{1/2}\ (L>1) - t_{1/2}(L<<1)$, the ratio of bimolecular recombination coefficient (B) with Langevin's one according to expression:

$$\frac{B}{B_L} = \frac{t_{tr}^2}{t_e(t_e + t_{tr})}\frac{1}{\alpha d} \qquad (5)$$

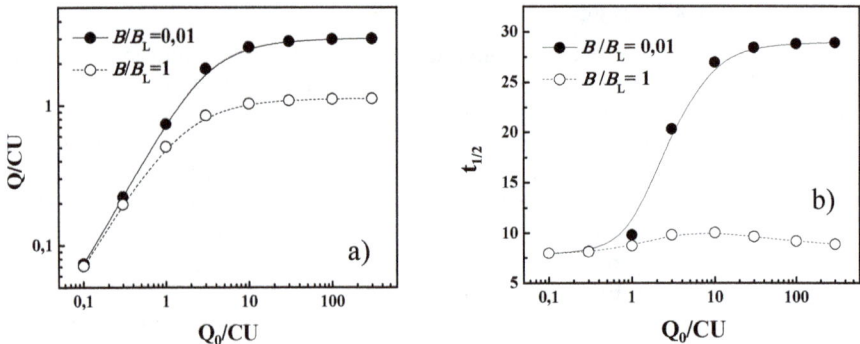

Fig. 3. Dependencies of amount of extracted charge Q (a) and of photocurrent halfwidth (b) on amount of photogenerated charge in case of Langevin and reduced bimolecular recombination

Here α is absorption coefficient. However, it is easer to measure the recombination coefficient using integral TOF when $RC > t_{tr}$ (Pivrikas et al, 2005). The examples of numerically modelled transients are demonstrated in Fig.2. Using this method the coefficient of bimolecular recombination is estimated as (Fig. 3):

$$B = \frac{edS}{t_e Q} \tag{6}$$

2.2 Charge carrier extraction by linearly increasing voltage (CELIV) method.

Method has the advantage that it is suitable for investigation of both high and low conductivity materials (Juška et al, 2000 a; Juška et al, 2004). After the triangular voltage pulse is connected to the sample electrodes in backward direction, the current caused by geometric capacitance of sample ($j(0)$) and conductivity current Δj are observed (Fig. 4).

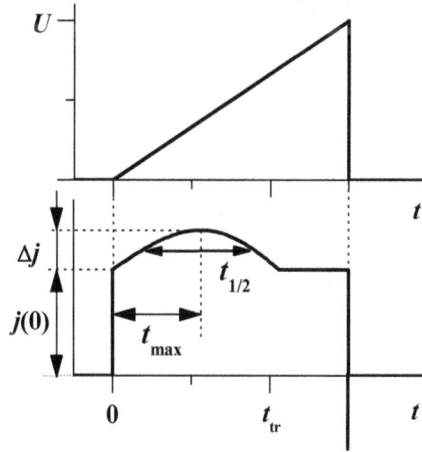

Fig. 4. Voltage pulse and current density observed by CELIV method

The measuring device is very simple: triangular pulse voltage generator and oscilloscope. Another advantage is that after triangular pulse of voltage is applied onto sample electrode, there is no initial, caused by capacitance, current peak, which disturb to monitor drift current in conductive materials.

The current transients were calculated by using standard solution method from continuity, current and Poisson equations in case when one of electrodes is blocking: Schottky or p-i barrier, or even special structure with isolating sublayer. From Poisson equation when density of equilibrium carriers is n_0 the extraction depth l ($0 \le l \le d$) is estimated:

$$\frac{Q(t)}{\varepsilon \varepsilon_0} = \frac{en_0 l(t)}{\varepsilon \varepsilon_0} = E(0,t) - E(d,t), \tag{7}$$

here Q is the amount of extracted charge, $E(0,t)$ and $E(d,t)$ are the magnitudes of electric field at the front and back electrodes correspondingly.
From the continuity equation:

$$\frac{dQ}{dt} = j_d = \sigma_0 E(d,t) ,$$
(8)

In case of linearly increasing voltage $U = A\,t$ and

$$\int_0^d E dx = At = E(d,t) \cdot d + \frac{E(0,t) - E(d,t)}{2} \cdot l(t) ,$$

the Rikati equation is obtained for $l(t)$

$$\frac{dl(t)}{dt} + \frac{\sigma}{2\varepsilon\varepsilon_0 d} l^2(t) = \frac{\mu At}{d} ,$$

Then the curent transient is

$$j(t) = \frac{\varepsilon\varepsilon_0 A}{d} + \frac{\sigma}{\mu}\left(1 - \frac{l(t)}{d}\right)\left(\frac{\mu At}{d} - \frac{\sigma}{2\varepsilon\varepsilon_0 d} \cdot l^2(t)\right) ,$$
(9)

The first component is caused by capacitance, and second one by conductivity.
When $\tau_\sigma = \varepsilon\varepsilon_0/\sigma >> t_{tr}$

$$j(t) = \frac{A}{d}\left[\varepsilon\varepsilon_0 + \sigma t\left(1 - \frac{\mu At^2}{2d^2}\right)\right], \text{ when } t < d\sqrt{\frac{2}{\mu A}} = t_{tr} ,$$
(10a)

$$j(t) = \frac{A}{d} \cdot \varepsilon\varepsilon_0 = j(0), \text{ when } t > t_{tr}$$
(10b)

From experimentally observed current transient the thickness of sample and/or dielectric permitivity may be estimated:

$$\frac{\varepsilon\varepsilon_0}{d} = \frac{j(0)}{A} ,$$
(11)

The dielectric relaxation time may be estimated as

$$\tau_\sigma = \frac{2}{3} \cdot t_{max} \frac{j(0)}{\Delta j} ,$$
(12)

The mobility of equilibrium charge carriers can be estimated as

$$\mu = \frac{2d^2}{3At_{max}^2} \text{ if } \Delta j \le j(0), \text{ i.e. } \tau_\sigma \ge t_{tr}$$

$$\mu = \frac{\tau_\sigma d^2}{At_{max}^3} \text{ if } \Delta j >> j(0), \text{ i.e. } \tau_\sigma << t_{tr}$$
(13)

The bulk conductivity of sample follows from:

$$\sigma_{bulk} = \varepsilon\varepsilon_0 \left.\frac{dj}{dt}\right|_{t=0} / j_0 .$$
(14)

The density of charge may be calculated from:

$$p_0 = \frac{2}{ed} \int_0^\infty \Delta j \, dt \quad . \tag{15}$$

In Fig. 5 there are demonstrated the results of modelling without trapping and with single trap level (Juška et al, 2000, b). For high and low A the modelling very well reproduces $t_{max}(A) \approx A^{-0.5}$ and $A^{-0.33}$, predicted by Eq. (13). When trapping is accounted then in both limiting cases the same expressions for t_{max} and Δj like without trapping are obtained, if one substitute μ by (μf), where f is the trapping factor or for single trap $f = \tau_C/(\tau_C + \tau_R)$, where τ_R is the release time.

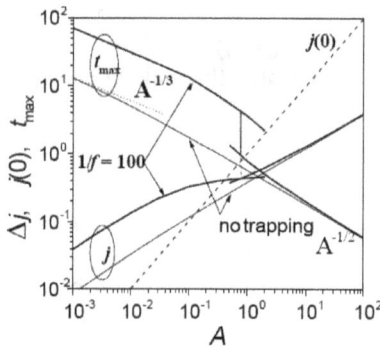

Fig. 5. Numerical modelling results of dependencies of Δj, t_{max}, $j(0)$ on A. $A = 1$, when $t_{tr} = \tau_\sigma$. Bold lines demonstrate dependencies when shallow trapping is accounted ($\tau_C = 1$, $\tau_R = 100$); lines correspond case when shallow trapping is absent. Density of current is normalized to magnitude of $j(0)$, when $A = 1$, and time is normalized to τ_σ

The basic measurable parameters of CELIV Δj, t_{max} depend on charge carrier interaction with trapping states, and this is reflected in dependencies of t_{max} and Δj (Fig. 6). Numerical modelling (Juška et al, 2000 b), taking into account energy distribution of trapping states as $N(E) \sim \exp(-E^2/2\delta^2)$, demonstrate that measurements of $\Delta j(A) \sim A^\beta$ and $t_{max}(A) \sim A^\gamma$ dependencies in various temperatures and electric fields, while choosing such A that $\Delta j \cong j(0)$, and estimating the rates of change as coefficients $\beta = \dfrac{d(\ln j)}{d(\ln A)}\bigg|_{\Delta j = j(0)}$ and

$\gamma = \dfrac{d(\ln t_{max})}{d(\ln A)}\bigg|_{\Delta j = j(0)}$, the nature of charge carrier interaction with trapping states can be cleared up, i.e., which charge carrier transport model is prevailing (Fig. 6):

1. if $\mu(F)$ dependence is caused by stochastic transport, then $(\beta - \gamma) = 1$, $(\beta + \gamma) < 0$;
2. if $\mu(E)$ dependence is caused by Poole-Frenkel type dependence of micromobility on electric field $(\mu \sim \exp(a\sqrt{E}))$ then $(\beta - \gamma) > 1$, and $(\beta + \gamma) < 1$, and the latter is independent or decreases with increasing a (T decreases);
3. if the characteristic release from trapping states time τ_R depends on electric field, i.e. $\tau_R \sim \exp(-b\sqrt{E})$ then, when b increases, $(\beta - \gamma) > 0$, and $(\beta + \gamma)$ increases or even changes the sign.

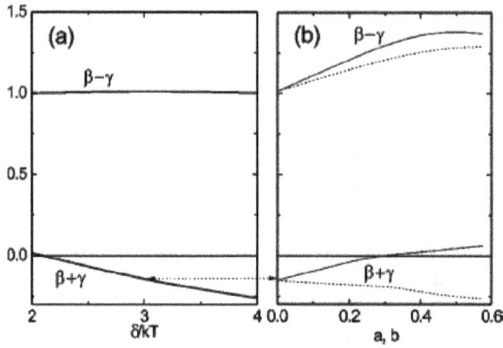

Fig. 6. Numerical modelling results of $(\beta + \gamma)$ and $(\beta - \gamma)$ dependencies on: (a) parameter δ/kT of Gaussian distribution of localized states; (b) Poole-Frenkel parameters a (doted line) and b (line) when $\delta/kT = 3$

2.3 Photo-CELIV method

Photo-CELIV method demonstrate even more opportunities where, by short pulse of light, photogenerated charge carriers are extracted by delayed (delay time t_{dU}) triangular pulse of voltage (Fig. 7) (Österbacka et al, 2004). Measurements of amount of extracted charge dependence on the delay time t_{dU} allow investigation of the relaxation of charge carrier density and mobility, independently. The latter are important in case of stochastic transport.

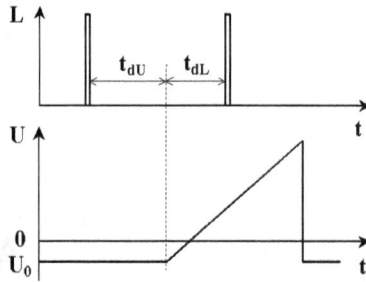

Fig. 7. Time chart of photo-CELIV method

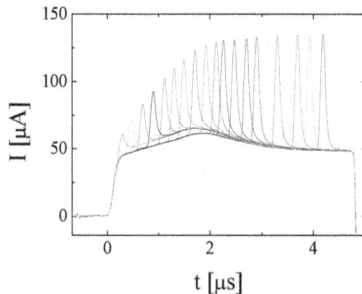

Fig. 8. Photocurrent transients of photo – CELIV for different t_{dL}

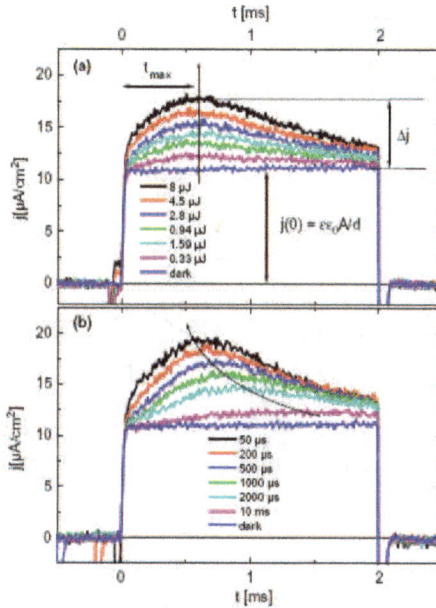

Fig. 9. Photocurrent transients of photo-CELIV at different intensity of pulse of light and fixed delay time (a), and at different delay time t_{dU} and fixed intensity of pulse of light (b) in RRa-PHT layer

2.4 Double injection current transient (DoI) method.

After the voltage is applied onto solar cell's electrodes in forward direction, the double injection current is observed. When the dielectric relaxation time is longer than charge carrier drift time ($\tau_\sigma = \varepsilon\varepsilon_0/\sigma \gg t_{tr}=d/\mu E$), in case of bimolecular Langevin recombination, the whole injected charge carriers recombine while moving through interelectrode distance, and the observed current transient matches space charge limited current transient in case of sum of mobilities of both sign carriers.

When the recombination is weaker, then, after the drift time (t_{sl}) of slower charge carriers, an amount of injected charge carriers and, at the same time, current increases till saturates, due to recombination. Thus, the dependence of saturated density of current on voltage is:

$$j = 2\sqrt{\frac{e\varepsilon\varepsilon_0\mu_n\mu_p\left(\mu_n + \mu_p\right)}{B}}\frac{U^2}{d^3} = 2\varepsilon\varepsilon_0\sqrt{\frac{B_L}{B}\mu_n\mu_p}\frac{U^2}{d^3} \cdot , \text{ when } \tau_\sigma \gg t_{tr}; \tag{16}$$

From the shape of current transient pulse it is possible evaluate whether recombination is of Langevin-type or weaker. In Fig. 10 there are shown measurable parameters, from which the transport and recombination values is estimated. The sum of mobilities of both sign charge carriers as

$$\mu_n + \mu_p = 0.8\frac{d^2}{Ut_{sc}} \tag{17}$$

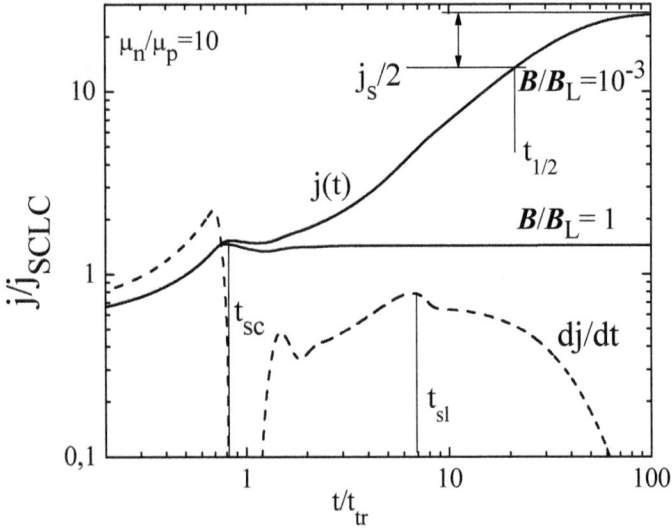

Fig. 10. Transients of double injection currents into dielectric in cases of Langevin and reduced bimolecular recombination

or in other case

$$\mu_n + \mu_p \cong \frac{j_{SCLC}d^3}{\varepsilon\varepsilon_0 U^2}.$$

(18)

The mobility of slower charge carriers

$$\mu_{sl} = 0.8\frac{d^2}{Ut_{sl}}.$$

(19)

The coefficient of bimolecular recombination

$$B = \frac{\ln 3}{2}e\left(\mu_n + \mu_p\right)\frac{U}{j_s t_{1/2}d}.$$

(20)

In case of plasma injection into semiconductor, when dielectric relaxation time is shorter than charge carrier transit time $(\tau_o = \varepsilon\varepsilon_0/\sigma \gg t_{tr}=d/\mu E)$, after ambipolar transit time of charge carriers $(t_a = t_m)$, an amount of injected plasma, and, thereby, the current, increases till, due to recombination, saturates.

$$j(t) = \begin{cases} \sigma E\left(1 - \frac{2}{3}\frac{t}{t_a}\right)^{-\frac{1}{2}}, & t < \frac{5}{6}t_a \\[2mm] \frac{3}{2}\sigma E + \left(j_s - \frac{3}{2}\sigma E\right)\tanh\left(B\Delta n_s\left(t - \frac{5}{6}t_a\right)\right), & t > \frac{5}{6}t_a \end{cases}$$

$$j_s = \frac{8}{9}e\sqrt{\frac{\left(\mu_p + \mu_n\right)\mu_p\mu_n\left(n_0 - p_0\right)}{B}} \cdot \frac{U^{3/2}}{d^2} = \frac{8}{9}\cdot\frac{U}{d}\varepsilon\varepsilon_0\sqrt{\frac{B_L}{B}}\cdot\frac{1}{\tau_\sigma t_a} \text{ , when } \tau_\sigma \ll t_a. \quad (21)$$

From the maximum of differential of current transient, using Eq. (22), the ambipolar mobility (μ_a) is estimated:

$$t_m\left(\frac{dj}{dt}\bigg|_{max}\right) = t_a = \frac{5}{6}\cdot\frac{d^2}{\mu_a U} . \quad (22)$$

Using Eq. (23), the coefficient of bimolecular recombination is:

$$B/B_L = 0.45\frac{\tau_\sigma t_m}{t_{1/2}^2} . \quad (23)$$

From shape of double injection current pulse the information about charge carrier trapping is obtained (Fig. 11) (Juška et al, 2008). During the trapping of the slower charges, after the transit time, through the interelectrode distance, of faster charge carriers, the space charge limited current is flowing through the sample till the whole trapping states are filled in by slower charge carriers ("hole trapping" in Fig. 11). When the trapping of faster charge carriers is dominating, the current is decreasing and begins to increase after trapping states are filled in ("electron trapping" in Fig. 11). Thus, the integration of current until time when the current starts to rise, allows evaluate density of trapping states.

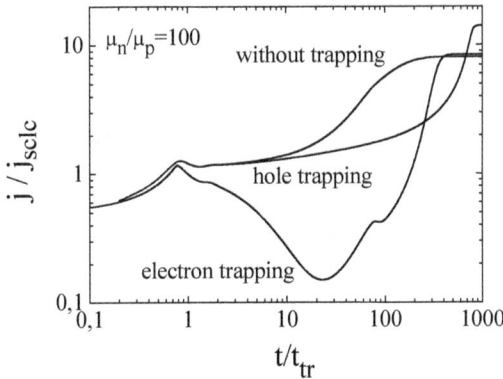

Fig. 11. Numerical modelling of double injection current transients when trapping is absent and when faster or slower charge carriers are trapped

The high capacitance of thin solar cells, immediately after application onto electrodes of rectangular pulse of voltage, causes high initial spikes of current, which complicates measurement and analysis of double injection current transients. To around the latter problem is possible by modification of DoI method, i.e. immediately after forward voltage pulse to apply the pulse of backward direction, and to measure the extraction current transient (Fig. 12) (Juška et al, 2006).

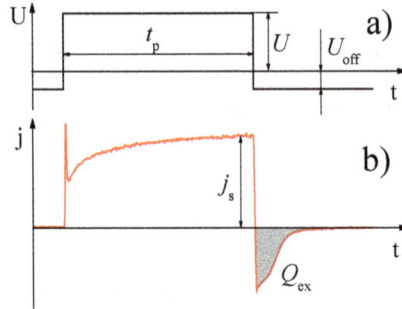

Fig. 12. Time charts of voltage pulse (a) and transients of double injection and extraction currents

The integral of extraction current gives an amount of extracted charge, from which's dependence on duration of injecting voltage (Fig. 13), the charge carrier mobility and bimolecular recombination coefficient can be estimated as:

$$B = \frac{\ln 3}{2} \frac{edS}{t_{1/2}Q_s}. \tag{24}$$

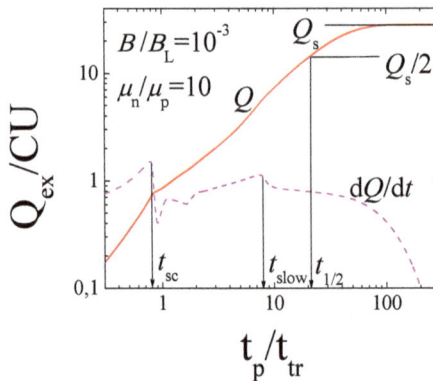

Fig. 13. Dependence of extracted charge on duration of injecting voltage pulse

3. Experimental results

3.1 Amorphous and microcrystalline silicon

Light-induced degradation of amorphous hydrogenated silicon (a-Si:H) is a serious problem of a-Si:H based photovoltaic solar cells. The most probable driving force for a-Si:H degradation is the energy (more than 1 eV) released during nonradiative bimolecular recombination of electron-hole pairs (which prevails at high light intensity) and that is why the discovery of mechanism of this recombination is of great importance.

For the study of bimolecular recombination coefficient (B) we have proposed the photoelectrical method (Juška et al, 1995), which is based on the measurement of extraction time (t_e) of the charge carrier reservoir using the space-charge limited photocurrent (SCLP) transient method. This method gives a possibility to estimate the monomolecular recombination time from the shape of the t_e dependence on the light intensity (L) and the bimolecular recombination coefficient B from the saturated value of t_e. These photoelectrical measurements demonstrated that the bimolecular recombination begins to prevail if charge carrier density is approximately 10^{17} cm^{-3}, and $B \cong 10^{-9}$ cm^3/s.

In a-Si:H layers it was observed the reduced bimolecular recombination, which, possibly, is reduced because electron and hole, immediately after photogeneration, are separated by internal random potential field. Fig. 14 demonstrates that the bimolecular recombination coefficient is lower in a-Si:H layers, which are deposited at high grow speeds (internal random field is greater), and that temperature dependence of B is stronger than that of high-quality amorphous silicon layers.

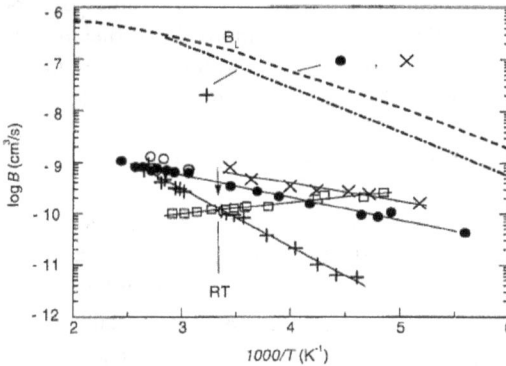

Fig. 14. Dependence of bimolecular recombination coefficient of electrons ($\bullet$) and holes ($\circ$) on temperature in: high grade a-Si:H ($\times$, $\bullet$, $\circ$), high deposition rate a-Si:H ($+$) and μc-Si:H ($\square$) layers. Temperature dependencies of Langevin recombination coefficient (B_L)

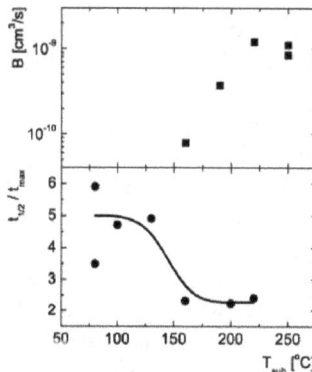

Fig. 15. Dependencies of bimolecular recombination coefficient B and dispersion parameter, estimated as $t_{1/2}/t_{max}$ from CELIV (Fig. 4), on substrate temperature during deposition of μc-Si:H layer

In similar way the internal random potential influences bimolecular recombination in microcrystalline hydrogenated silicon (μc-Si:H). The temperature of substrate during deposition of μc-Si:H strongly influences the magnitude of internal random potential, and, through the latter, influences dispersion of charge carrier transport. Thus, decreasing of the substrate temperature leads to increase of dispersion of charge carrier transport, but decreases coefficient of bimolecular recombination (Fig. 15).

3.2 π-conjugated polymers

Recently, the great opportunity to create enough effective, large area and low-cost organic solar cells (OSC) increased interest in π-conjugated polymers, but also has raised several problems. First of all, in disordered materials, which include π-conjugated polymers, the mobility of charge carriers, due to hopping, is low ($\mu \ll 1$ cm²/Vs) in comparison with crystalline materials, and the mean hopping distance of charge carriers is shorter than Coulomb radius. In turn, the low charge carriers jump distance results in a low photogeneration quantum efficiency and conditions the diffusion-controlled, Langevin type charge carrier recombination. Latter is caused by, the diffusion and Coulomb inter-traction field controlled, meeting probability of electron and hole in space. On the other hand, in order to OSC current density would be comparable with the crystalline semiconductors the density of photogenerated charge carriers should be much higher than in the crystalline solar cells.

The density of charge carriers, due to bimolecular recombination, causes small their lifetime because $\tau = (Bn_f)^{-1}$. However, for effective OSC, it is necessary that the lifetime of charge carriers should be higher than their drift time through the interelectrode distance in intrinsic electric field, i.e. $\tau > t_{tr}^i = d^2 / \mu U_i$ (U_i is intrinsic potential). Thus, for higher than 5% efficiency of OSC, when open circuit voltage is ~ 0.5 V, thickness of sample is 300 nm, it is necessary that density of photocurrent will be higher than 15 mA/cm², and $\mu B_L / B > 5 \times 10^{-3}$ cm²/Vs. Thus, the bimolecular recombination limits efficiency of organic solar cell in region of high intensity light, and ratio of bimolecular recombination coefficient with Langevin's one allows evaluate effectiveness of materials and structures.

As a model material for investigation of features of bimolecular recombination was chosen π-conjugated polymer RRa PHT. In RRa PHT layer the TOF current transients were nondispersive at low intensity of light pulses. With increase of intensity of the light pulse, the shape of photocurrent transient changes to the classic SCLC kinetics, and, at a very high intensity of light, its shape stopped to change (Fig. 16). An amount of extracted charge linearly increased with intensity of light and saturated in the region of high light intensity, when $Q_e / CU_0 = 1$. Such saturation of the $Q_e(L)$ dependence is the consequence of Langevin recombination (See Eq. (4)).

To assess the coefficient of Langevin recombination, it is necessary to know the charge carriers mobility, when the electric field is zero, because in the depth of photogeneration the electric field is shielded by the carriers. By measuring the hole mobility dependence on electric field strength and by extrapolation to $E = 0$, the hole mobility was evaluated as $\mu_p(E = 0) = 6.5 \times 10^{-6}$ cm²/Vs, and, considering that the mobility of electrons at least by one order lower than of the holes, $B_L = 4 \times 10^{-12}$ cm³/s was estimated.

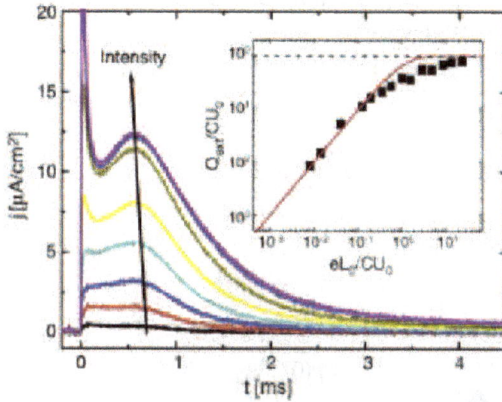

Fig. 16. TOF transients of photocurrent for different intensity pulse of light. Thickness of RRa PHT layer is 8 µm, $E = 10^5$ V/cm

For direct measurements of photogenerated charge carrier density and mobility relaxation on time, the photo-CELIV method was used. From the duration t_{max} (see Fig. 4) the mobility of charge carriers and from integral of conductivity current ($\frac{1}{e}\int_0^\infty \Delta j dt$, here Δj is density of conductivity current) the density of charge carriers (p) at given t_{dU} (see Fig. 7) are estimated. The presence in the structure of intrinsic electric field has been compensated by offset voltage. Possible inaccuracy of this method can be caused by spatially distributed intrinsic electric field, which separates photogenerated charge carriers, thereby, decreasing recombination.

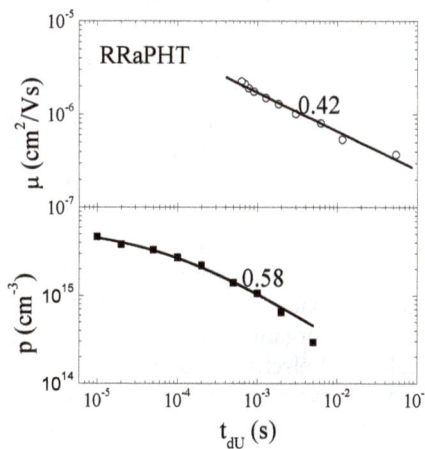

Fig. 17. Charge carrier mobility and density dependencies on t_{dU}. Solid line is $p(t)$ results according $B(t) = e\mu(t) / \varepsilon\varepsilon_0$ in case of Langevin recombination

In Fig. 17 the typical for RRa PHT mobility and density dependencies of hole on time are demonstrated. Fitting of mobility relaxation as $\mu = at^{-0.42}$ gives that Langevin recombination coefficient changes with time too, i.e. $B(t) = e\mu(t)/\varepsilon\varepsilon_0$. Thus, charge carrier density follows expression

$$p(t) = \frac{p(0)}{1 + p(0)B \int\limits_0^t \mu(t)dt}. \tag{25}$$

Therefore

$$p(t) = \left(\frac{1}{p(0)} + \frac{e}{\varepsilon\varepsilon_0} \cdot \frac{at^{0.58}}{0.58} \right)^{-1}. \tag{26}$$

which is shown by solid line in Fig. 17. The coincidence of experimental results and theory confirms that, in low mobility organic material, bimolecular recombination is of Langevin-type. So, the same result has been obtained from the saturation of SCLC transients with intensity of light (Pivrikas et al, 2005 [5]).

3.3 Recombination in conjugated polymer/fullerene bulk heterojunction solar cells

One of possibilities reduce bimolecular recombination is to make junction of two organic material layers, in one of which are mobile the electrons and in another one the holes. The excitons, immediately after photoexcitation, are destroyed by electric field of heterojunction and separated electrons and holes are moving each of its transport material to sample electrodes. However, in organic polymers the diffusion distance not exceed 100 nm. Thus, an efficiency of such solar cell would be low because the thickness of solar cell would be approximately 100 nm and absorption of light weak. From Fig. 18a follows, that, oppositely to MEH PPV (poly(2-methoxy-5-(2'-ethylhexykoxy)-1,4-phenylenevinylene) layer, the heterojunction of MEH PPV/perilene more effectively separates photogenerated pairs, i.e. the charge carrier reservoir is created and an amount of collected charge approximately twice exceeds CU_0 (Fig. 18b). Supporting the latter experimental result, the numerical modelling, taking into account the Langevin recombination, demonstrates that, in case of bulk absorption of light and of small resistor, causing extraction current, an amount of extracted charge can exceed CU a few times, too. Thus, the obtained experimental results do not deny the Langevin recombination in heterojunction.

Another charge carriers bimolecular recombination reduction method has been identified investigating a-Si:H and μc-Si:H layers: separate photogenerated charge carriers by an internal random field in space, so, that they move towards the electrodes in different ways. This method has been used for organic semiconductor structures: layer cast mixing transporting materials of holes and electrons. Such a bulk heterojunction blends allow to expect a significant reduction of bimolecular recombination, as in the bulk of samples created excitons are in the vicinity to heterojunctions. When they disintegrate, resulting electrons and holes moving towards each of its material to the contrary of the electrodes, i.e. separated in space.

Experimentally there were investigated bulk heterojunctions of various organic polymers with PCBM.

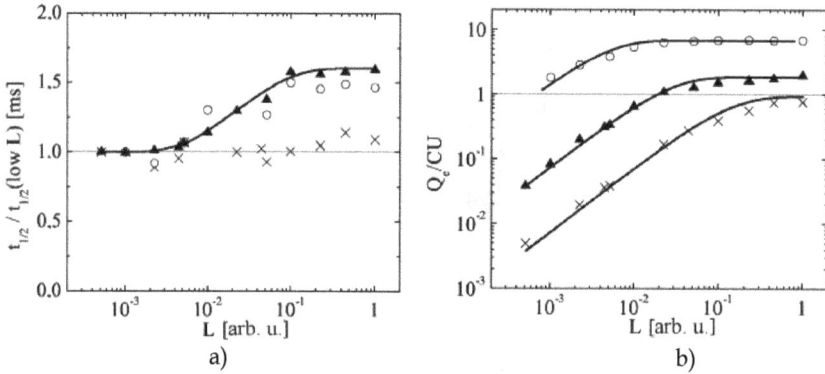

Fig. 18. Dependencies of TOF current transient halfwidth (a) and of collected charge (b) on the intensity of light pulse: × - MEH PPV, ▲ – perilene/MEH PPV junction, O - MEH PPV:PCBM blend

Time-dependent mobility and recombination in the blend of poly[2-methoxy-5-(3,7-dimethyloctyloxy)-phenylene vinylene] (MDMO-PPV) and 1-(3-methoxycarbonyl)propyl-1-phenyl-(6,6)-C_{61} (PCBM) is studied simultaneously using the photoinduced charge carrier extraction by linearly increasing voltage technique (Mozer et al, 2005).

Photo-CELIV transient at various delay times, light intensities and applied voltages have been recorded, and the charge carrier mobility and lifetime simultaneously studied. It is found that, shortly after photoexcitation, both the charge mobility and the recombination are time-dependent (dispersive) processes, which is attributed to the initial relaxation of the charge carriers towards the tails states of the density of states distribution. The results confirm that the recombination dynamics within the studied μs - ms time scale is a thermally activated process rather than a temperature independent tunneling. The obtained time-dependent mobility values are used to directly describe the recombination dynamics (see Fig. 19 and Fig. 20). Density decay of charge carriers fitted according to Eq. (25).

Therefore results suggest that the recombination dynamics is nearly Langevin-type, i.e. controlled by diffusion of the charge carriers towards each other.

3.4 Recombination in P3HT:PCBM Bulk Heterojunction Organic Solar Cells

In poly(3-hexylthiophene): 1-(3-methoxycarbonyl) propyl-1-phenyl[6,6]C_{61} (P3HT:PCBM) bulk heterojunction solar cells, a reduction of the Langevin recombination is commonly observed after thermal treatment. This treatment has been shown to modify significantly the nanomorphology of the photoactive composite, inducing a crystallization of both the donor and the acceptor phases (Pivrikas et al, 2007). In Fig. 21 the experimentally measured results using integral TOF SCLC regime are presented. By comparing experimentally measured bimolecular recombination coefficient B with calculated Langevin recombination coefficient B_L, it was shown that $B/B_L \cong 10^{-3}$.

According to (Adriaenssens et al, 1997), if the reduction of bimolecular recombination is caused by random potential, the bimolecular recombination has follow $B \cong B_L \exp(-\Delta E/kT)$ dependence on temperature (here ΔE is mean random potential energy). Thus, the activation energy of B has to be higher than one of B_L, while experimentally it is obtained an opposite result. In case if bimolecular recombination is caused by tunnelling, the B/B_L ratio will

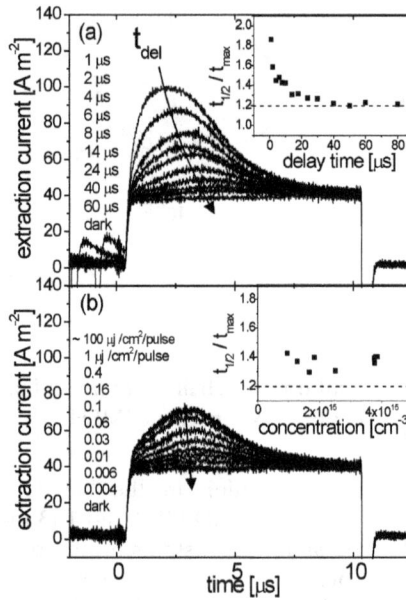

Fig. 19. Photo-CELIV transients recorded at 300 K at (a) various delay times at fixed light intensity; (b) varying illumination intensities attenuated using optical density filters at fixed 5 μs delay time. The voltage rise speed A was 4 V/10 μs. The insets show the calculated dispersion parameters $t_{1/2}$ to t_{max} versus delay time and the concentration of the extracted charge carriers, respectively.

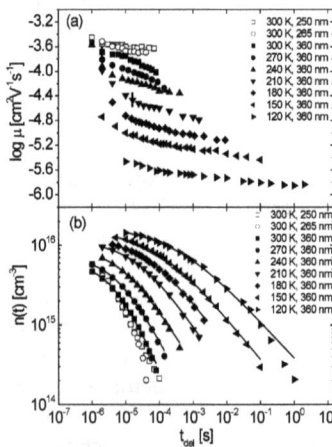

Fig. 20. Mobility (a) and the density (b) of extracted charge carriers versus the delay time for samples with different active layer thickness. Charge carrier mobility and density measured for the 360 nm device. Density relaxation of charge carriers fitted according to Eq. (25)

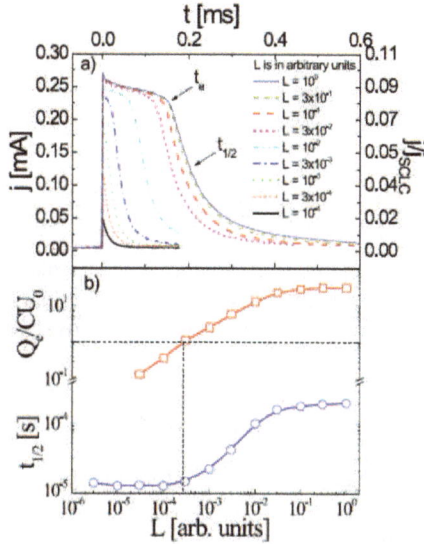

Fig. 21. Dependencies of integral TOF photocurrent transients (a), and of Q_e/CU_0 and $t_{1/2}$ on intensity of light pulse in RRP3HT:PCBM bulk heterojunction. Horizontal dotted line (b) corresponds $Q_e = CU_0$

demonstrate strong dependence on electric field. However, as is obvious from experimental results, this is not a case. The increasing of random potential also did not cause reduction of B/B_L ratio. In blend of segmented electron and hole transporting materials the meeting of charge carriers of opposite sign may be limited by charge carriers of lower mobility. However, this is insufficient to explain such big reduction of bimolecular recombination in RRP3HT:PCBM blend. This can be explained by, that the interface between polymer and acceptor materials, decreases Coulomb interaction, which suppress gemini and bimolecular recombination as it was proposed in (Arkhipov Heremans & Bässler, 2003).

Furthermore, double injection current transients (DoI) and photo-CELIV measurements revealed, that the reduced B depends on the charge carrier density as in the case of Auger recombination (Juška et al, 2008). The same conclusion followed from transient photo voltage and transient photo absorption spectroscopy experiments (Shuttle et al, 2008). In the recent transient absorption spectroscopy experiments (Nelson, 2003) it was suggested that this type of relaxation is caused by a stochastic transport attributed to an exponential tail of localized states. However photo-CELIV and TOF experiments showed that the photocurrent relaxation is caused by the charge carrier's recombination (Pivrikas et al, 2005).

In this work, we are demonstrating that the reduction of B and its dependence on charge carriers density is caused by the two dimensional Langevin recombination (Juška et al, 2009).

Sirringhaus (Sirringhaus et al, 1999) showed that the mobility across and along the lamellar structure differs more than 100 times, which led to the fact that the recombination of charge carriers is mainly taking place in the two-dimensional lamellar structure. When spacing between lamellas $l << r_m$, r_m is determined by $\pi r_m^2 l = 1/n$. Then the recombination probability

$$f_{2D} = \frac{1}{t_m} = \frac{3\sqrt{\pi}}{4} \cdot \frac{e(\mu_n + \mu_p)}{\varepsilon\varepsilon_0}(l \cdot n)^{3/2} = \gamma_{2D}n^{3/2} \, . \tag{27}$$

where γ_{2D} is 2D recombination parameter. Hence, in two dimensional case, the bimolecular recombination coefficient will be reduced in comparison with one of the three dimensional case as

$$\frac{B_{2D}}{B_{3D}} = \frac{3\sqrt{\pi}}{4}l^{3/2}n^{1/2} \, .$$

For RRP3HT $l \cong 1.6$ nm (Sirringhaus et al, 1999) and, for example, when $n = 10^{16}$cm^{-3}, β_{2D}/β_{3D} = 6×10^{-3}, and that is close to experimental results of Ref. (Pivrikas et al, 2007).
In the 2D recombination case, using Eq. (27), the equation governing the decay of the charge carriers is:

$$\frac{dn}{dt} = G - \gamma_{2D}n^{5/2} \, . \tag{28}$$

where G is the rate of the photogeneration or double injection. Similar dependencies were observed experimentally: $dn/dt \propto n^{2/6}$ (Shuttle et al, 2008). According to Eq. (28), after excitation by short pulse of light, the decay of the density of charge carriers is described as:

$$n(t) = \left(n_0^{-3/2} + \frac{3}{2}\gamma_{2D}t \right)^{-2/3} \propto t^{-2/3} \Bigg|_{n_0 \to \infty} \, , \tag{29}$$

here n_0 is initial density of photogenerated charge carriers. The similar dependence is observed using photo-CELIV technique (Fig.22).
In the case of 3D Langevin recombination $n(t) \propto t^{-1}$. It is worth to notice, that the slower than $n(t) \propto t^{-1}$ dependence can be observed due to the mobility dependence on time (stochastic transport), as it was shown in regiorandom poly(3-hexylthiophene) (Pivrikas et al, 2007). However, it is established by photo-CELIV that the mobility does not depend on the delay time after excitation (Fig. 22). That is why this explanation is not valid for the RRP3HT:PCBM blends. Another technique, which allows investigation of recombination process, is DoI current transient technique (Juška et al, 2007).

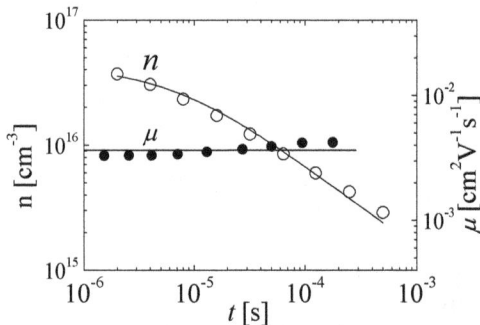

Fig. 22. Measured using photo-CELIV dependencies of mobility and density of electrons on time in annealed RR-P3HT:PCBM bulk heterojunction.

In the case of 3D Langevin recombination volt-ampere characteristics and DoI current transients corresponds to the sum of the space charge limited currents of electrons and holes, because the injected charge carriers will recombine completely within the interelectrode distance. In Fig. 23a numerically modelled DoI current transients are shown for the both 3D and 2D Langevin recombination cases for different distance l and ratios between fast and slow charge carriers mobilities μ_f / μ_{sl}.

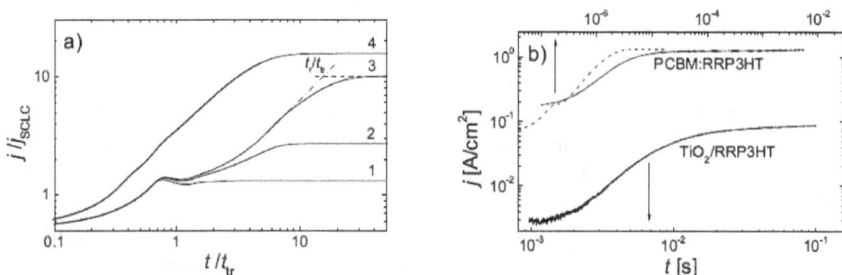

Fig. 23. Numerically modelled DoI current transients: a) in the case of 3D Langevin recombination (1) and 2D Langevin recombination for [(2) $l = 10$ nm, (3) $l = 1$ nm and μ_f / μ_{sl} = 10, (4) $l = 1$ nm and μ_f / μ_{sl} = 1]; both the time scale and current are normalized to transit time t_{tr} and SCLC of faster charge carriers, respectively. (b) DoI current transients: solid lines – measurements of RRP3HT:PCBM bulk heterojunction ($d = 1.4$ μm, $U = 9$ V) and TiO$_2$/RRP3HT ($d = 0.6$ μm, $U = 4$ V) structure; dashed line - numerically modelled DoI current transient for RRP3HT:PCBM structure ($\mu_n = 10^{-2}$ cm^2/Vs, $\mu_p = 2.5 \times 10^{-3}$ cm^2/Vs, $l = 1.6$ nm)

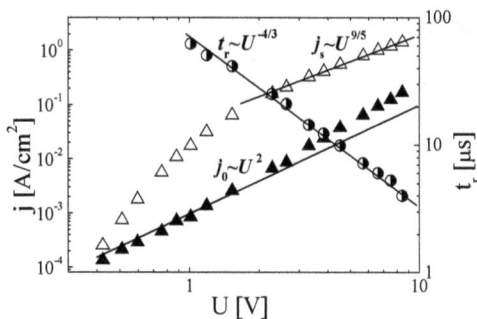

Fig. 24. The dependencies of initial current j_0, saturation current j_s and recombination time t_r on voltage in the case of double injection.

In the case of 2D Langevin recombination the current-voltage characteristics can be obtained from Eq. (16) in the same way as in Ref. (Juška et al, 2006): $j \propto U^{9/5} / d^{13/5}$, while the saturation time of the DoI current transient is $t_r \propto U^{-6/5}$. The observed experimental results (Juška et al, 2007) are very close to these dependencies.

The comparison of DoI current transients obtained experimentally and numerically modelled using $\mu_n = 10^{-2}$ cm^2/Vs, $\mu_p = 2.5 \times 10^{-3}$ cm^2/Vs and $l = 1.6$ nm (obtained from X-rays studies (Sirringhause et al, 1999) of RRP3HT:PCBM blend is presented in Fig. 23b. The ratios between initial and stationary currents are in good agreement with experiment giving the same l value, and the discrepancy in current rise times could be explained by not taken into account dispersion of the charge carrier transport.

So, the observed increase of DoI current in RRP3HT:PCBM bulk heterojunction and TiO$_2$/RRP3HT structures and good fit of numerical modelling with experimental data in RRP3HT:PCBM, proves that recombination takes place in RRP3HT. The slower rise of current in TiO$_2$/RRP3HT structure is caused by the lower electron mobility and deep trapping (Juška et al, 2008).

By the integral mode TOF method, where the RC time constant of the measurement setup is much larger than the transit time of the charge carriers ($RC \gg t_{tr}$) we can determine the 2D recombination parameter γ_{2D} in lamellas and its temperature dependencies in more convenient (Pivrikas et al, 2005) and straightforward way because it is independent on material's parameters. In the case of 3D Langevin recombination, the current transient saturates as a function of light-intensity and the amount of extracted charge slightly exceeds CU (when $\alpha d \gg 1$; $Q_{ex} = CU$, therefore $t_{ex} = 0$). In the case of 2D Langevin recombination, the charge carrier extraction time t_{ex}, when collected charge saturates with light intensity, is estimated in the similar way as in the case of reduced bimolecular recombination (Juška et al, 1995):

$$t_{ex} = \left(\frac{2}{3\gamma_{2D}} \right)^{2/5} \left(\frac{ed}{j_{ex}} \right)^{3/5} \propto j_{ex}^{-3/5} , \qquad (30)$$

where j_{ex} is extraction current, which could be varied by changing loading resistor or applied voltage. In the case of the reduced bimolecular recombination $t_{ex} \propto j_{ex}^{-1/2}$.

In Fig. 25b the extraction time as a function of the density of extraction current in different structures containing RRP3HT is shown. Since t_{ex} shows the same dependence on the extraction current density j_{ex}, it can be concluded that the recombination is taking place in RRP3HT and it is governed by the 2D Langevin recombination.

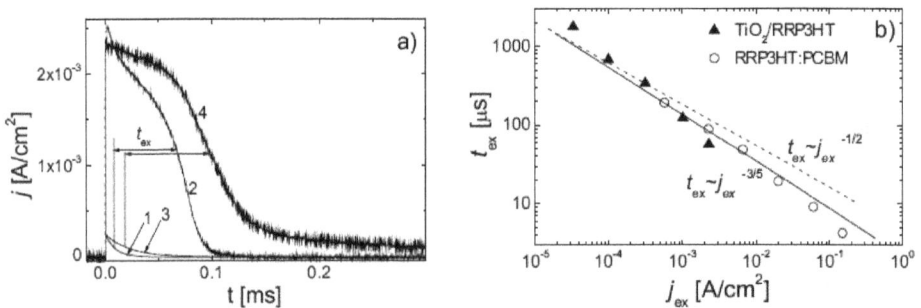

Fig. 25. Current transients of charge carrier extraction (a) observed by integral TOF: small charge drift current (1, 3) and transient of saturated on light intensity photocurrent (2, 4) in TiO$_2$/RRP3HT structures and RRP3HT:PCBM bulk heterojunction, respectively. Measurement of extraction time is indicated. Dependencies of extraction time (b) on the extraction current density in TiO$_2$/RRP3HT structures and RRP3HT/PCBM bulk heterojunction

4. Conclusion

In this work there are demonstrated the methods of investigation of charge carrier recombination in organic solar cells, where stochastic transport of charge carriers complicates interpretation of experimental results: charge carriers extraction by linearly increasing voltage (CELIV), which allows independently measure relaxation of density and mobility of photoexcited charge carriers; double injection current transient, which is additional method of investigation of charge carrier recombination and, which allows to measure dependence of recombination coefficient on electric field; integral time of flight (SCLC), using of which allows easily estimate the temperature dependence of recombination coefficient.

Experimentally it is shown, that the decay of the density of photogenerated charge carriers in the blend of MDMO-PPV:PCBM is of 3D Langevin-type, which is typical for organic materials, and in annealed samples of RRP3HT and bulk heterojunction solar cells of RRP3HT:PCBM it is of 2D Langevin-type recombination in the lamellar structure.

5. Referencies

Adriaenssens, G. J. & Arhipov, V. I. Non-Langevin recombination in disordered materials with random potential distributions. *Solid State Communications*, Vol. 103, Issue 9 (September 1997) 541- 543. ISSN 0038-1098.

Arkhipov, V. I.; Heremans, P. & Bässler, H., Why is exciton dissociation so efficient at the interface between a conjugated polymer and an electron acceptor? *Applied Physics Letters*, Vol. 82, Issue 25 (June 2003) 4605 1 - 3, ISSN 0003 6951.

Juška, G. ; Viliūnas, M.; Arlauskas, K.; Kočka, J. Space-charge-limited photocurrent transients: The influence of bimolecular recombination. *Physical Review B*, Vol. 51, No. 23 (June 1995) 16668 – 16676, ISSN 1098-0121.

Juška, G.; Viliūnas, M.; Arlauskas, K.; Stuchlik, J. & J. Kočka. Ultrafast Charge Carrier Recombination in a-Si:H and μc-Si:H. *Physica status solidi (a)*, Vol. 171, No. 2 (February 1999) 539 - 547, ISSN 1682-6300.

Juška, G.; Arlauskas, ; Viliūnas, M. & Kočka, J.. Extraction Current Transients: new method of study of charge transport in microcrystalline silicon. *Physical Review Letters*, Vol. 84, No. 21, (May 2000) 4946-4949, ISSN 0031–9007, a.

Juška, G.; Arlauskas, K.; Viliūnas, M.; Genevičius, K.; Österbacka, R. & Stubb, H. Charge transport in π-conjugated polymers from extraction current transients. *Physical Review B*, Vol. 62, No. 24 (December 2000) 16235-16238, ISSN 1098–0121, b.

Juška, G.; Genevičius, K.; Arlauskas, K.; Österbacka, R. & Stubb, H.. Features of charge carrier concentration and mobility in π-conjugated polymers. *Macromolecular Symposia*, Vol. 212, No. 1 (May 2004) 209-217, ISSN 1022–1360.

Juška, G.; Arlauskas, K.; Sliaužys, G.; Pivrikas, A.; Mozer, A. J.; Sariciftci, N. S.; Scharber, M. & Österbacka, R. Double injection as a technique to study charge carrier transport and recombination in bulk-heterojunction solar cells. *Applied Physics Letters*, Vol. 87, No. 22 (November 2005) 222110 1 – 3, ISSN 0003–6951.

Juška, G.; Sliaužys, G.; Genevičius, K.; Arlauskas, K.; Pivrikas, A.; Scharber, M.; Dennler, G.; Sariciftci, N. S. & Österbacka R. Charge-carrier transport and recombination in thin insulating films studied via extraction of injected plasma. *Physical Review B*, Vol. 74, No. 11 (September 2006) 115314 1-5, ISSN 1098-0121.

Juška, G.; Genevičius, K.; Sliaužys, G.; Pivrikas, A.; Scharber, M. & Österbacka, R. Double-injection current transients as a way of measuring transport in insulating organic films. *Journal of Applied Physics*, Vol. 101, No. 11 (June 2007) 114505 1-5, ISSN 0021-8979.

Juška, G.; Genevičius, K.; Sliaužys, G.; Nekrašas, N. & Österbacka, R.. Double injection in organic bulk-heterojunction. *Journal of Non-Crystalline Solids*, Vol. 354, Issues 19-25 (May 2008) 2858-2861, ISSN 0022-3093.

Juška, G.; Genevičius, K.; Nekrašas, N.; Sliaužys, G. & Österbacka, R.. Two dimensional Langevin recombination in regioregular poly(3-hexylthiophene), *Applied Physics Letters*, Vol. 95, No. 1 (July 2009), 013303 1-3, ISSN 0003-6951.

Mozer, A. J.; Dennler, G.; Sariciftci, N. S.; Westerling, M.; Pivrikas, A.; Österbacka, R. & Juska, G. Time-dependent mobility and recombination of the photoinduced charge carriers in conjugated polymer/fullerene bulk heterojunction solar cells, *Physical Review B*, Vol. 72, No. 3 (July 2005), 035217 1-10, ISSN 1098-0121.

Nelson, J. Diffusion-limited recombination in polymer-fullerene blends and its influence on photocurrent collection, *Physical Review B*, Vol. 67, No. 15 (April 2003) 155209 1-10, ISSN 1098-0121.

Österbacka, R.; Pivrikas, A.; Juška, G.; Genevičius, K.; Arlauskas, K. & Stubb, H. Mobility and density relaxation of photogenerated charge carriers in organic materials. *Current Applied Physics*, Vol. 4, No. 5 (August 2004) 534-538, ISSN 1567-1739.

Pivrikas, A.; Juška, G.; Österbacka, R.; Westerling, M.; Viliūnas, M.; Arlauskas, K. & Stubb, H. Langevin recombination and space-charge-perturbed current transients in regiorandom poly(3-hexylthiophene). *Physical Review B*, Vol. 71, No. 12, (March 2005) 125205 1-5, ISSN 1098-0121.

Pivrikas, A.; Juška, G.; ; Mozer, A. J.; Scharber, M.; Arlauskas, K.; Sariciftci, N. S.; Stubb, H. & Österbacka, R. Bimolecular recombination coefficient as a sensitive testing parameter for low-mobility solar-cell materials. *Physical Review Letters*, Vol. 94, No. 17, (May 2005) 176806 1 - 4, ISSN 0031-9007.

Pivrikas, A.; Sariciftci, N. S.; Juška, G. & Österbacka, R. A review of charge transport and recombination in polymer/fullerene organic solar cells. *Progress in Photovoltaics:. Research and Applications* , Vol. 15 (July 2007) 677-696, ISSN 1062-7995.

Sirringhaus, H.; Brown, P. J.; Friend, R. H.; Nielsen, M. M.; Bechgaard, K.; Langeveld-Voss, B. M. W.; Spiering, A. J. H.; Janssen, R. A. J.; Meljer, E. W.; Herwig, P. & de Leeuw, D. M. Two dimensional charge transport in self-organized, high mobility conjugated polymers. *Nature*, Vol. 401 (October 1999), 685 - 688, ISSN 0028-0836.

Shuttle, G.; O'Regan, B.; Ballantyne, A. M.; Nelson, J.; Bradley, D. D. C.; de Mello, J. & Durrant, J. R. Experimental determination of the rate law for charge carrier decay in a polythiophene: Fullerene solar cell. *Applied Physics Letters*, Vol. 92, No. 9, (March 2008), 093311 1 - 3, ISSN 0003-6951.

7

Uses of Concentrated Solar Energy in Materials Science

Gemma Herranz and Gloria P. Rodríguez
University of Castilla La Mancha. ETSII.
Metallic Materials Group
Avda. Camilo José Cela s/n. 13071. Ciudad Real.
Spain

1. Introduction

In recent decades tremendous advances have been made in the development of new materials capable of working under increasingly extreme conditions. This advance is linked to the development of Materials Surface Engineering. The utilisation of techniques based on high density energy beams (laser, plasma, electron beam or arc lamps) in surface modification and metallic material treatment allow for the creation of non-equilibrium microstructures which can be used to manufacture materials with higher resistance to corrosion, high temperature oxidation and wear, among other properties.

These techniques, despite their multiple possibilities, have one inconvenient property in common: their low overall energy efficiency. While it is true that the energy density obtained through a laser is three to four magnitudes greater than that which is obtained by solar energy concentration facilities, Flamant (Flamant et al. 1999) have carried out a comparison of the overall energy and the capital costs of laser, plasma and solar systems and came to the conclusion that solar concentrating systems appear to offer some unique opportunities for high temperature transformation and synthesis of materials from both the technical and economic points of view.

It is important to bear in mind that the use of this energy could lower the cost of high temperature experiments. Combined with the wide array of superficial modifications that can be carried out at solar facilities, there are numerous other advantages to using this energy source. The growing (and increasingly necessary) trend towards the use of renewable clean energy sources, which do not contribute to the progressive deterioration of the environment, is one compelling argument. Solar furnaces are also excellent research tools for increasing scientific knowledge about the mechanisms involved in the processes generated at high temperatures under non-equilibrium conditions. If, in addition, the solar concentration is carried out using a Fresnel lens, several other positive factors come into play: facility costs are lowered, adjustments and modifications are easy to carry out, overall costs are kept low, and the structure is easy to build, which makes the use of this kind of lens highly attractive for research, given its possible industrial applications.

These are the reasons that justify the scientific community's growing interest in researching the possible uses of highly concentrated solar energy in the field of materials. But this interest is not new. At the end of the 18th century, Lavoisier (Garg, 1987) constructed a concentrator based on a lens system designed to achieve the melting point temperature for platinum (1773°C). But it was not until the twentieth century that the full range of possibilities of this energy source and its applications to the processing and modification of materials started to be explored in depth. The first great inventor was Felix Trombe who transformed German parabolic searchlights used for anti-aerial defence during WW II into a solar concentrator. Using this device he was able to obtain the high temperatures needed to carry out various chemical and metallurgic experiments involving the fusion and purification of ceramics (Chaudron 1973). In 1949 he was able to melt brass resting in the focal area of a double reflection solar furnace which he constructed using a heliostat or flat mirror and a parabolic concentrator (50kW Solar Furnace of Mont-Louis, France). But his greatest achievement was the construction of the largest solar furnace that currently exists in the world, which can generate 100kW of power. The "Felix Trombe Solar Furnace Centre" is part of the Institute of Processes, Materials and Solar Energy (PROMES-CNRS) and is a leader in research on materials and processes.

Another of the main figures in the use of solar energy in the materials field and specifically in the treatment and surface modification of metallic materials is Prof. A.J. Vázquez of CENIM-CSIC. His research in this field started at the beginning of the 1990's, using the facilities at the Almería Solar Plant (Vazquez & Damborenea, 1990). His role in encouraging different research groups carrying out work in material science to experiment with this new solar technology has also been very important.

Our group's main focus at the ETSII-UCLM involved using concentrated solar energy (CSE) from a Fresnel lens to propose new sintering processes and surface modifications of metallic components. The aim was to increase the resistance of metallic materials (mainly ferrous and titanium alloys) to wear, corrosion and oxidation at high temperatures.

The initial studies with CSE at the ETSII-UCLM involved characterising a Fresnel lens with a diameter of 900 mm, for its use as a solar concentrator (Ferriere et al. 2004). The characterisation indicated that the lens concentrated direct solar radiation by 2644 times, which meant that on a clear day with an irradiance of $1kW/m^2$ the density of the focal area would be 264.4 W/cm^2 (Figure 1). This value is much lower than this obtained with other techniques based on high density beams, but is sufficiently high to carry out a large number of processes on the materials, and even a fusion of their surfaces.

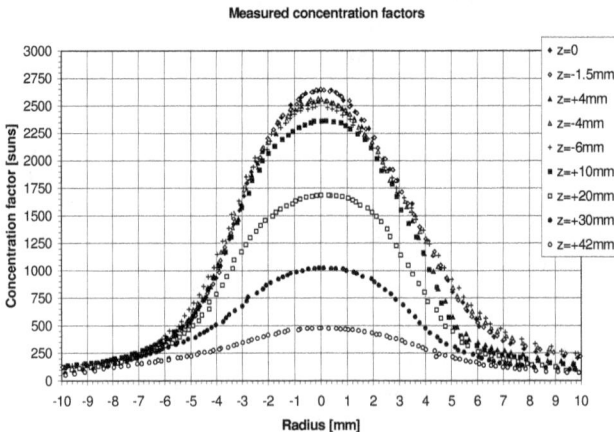

Fig. 1. Concentration factor of the Fresnel lens

The investigations carried out to date include processes involving the sintering of metallic alloys, surface treatment of steel and cast irons, cladding of stainless steel and intermetallic compound, high temperature nitriding of titanium alloys and NiAl intermetallic coating processing through a SHS reaction (Self-propagating high temperature synthesis). This research has been carried out in European and national programmes for Access to Large-Scale Facilities which allowed us to collaborate with the groups of A. J. Vázquez (CENIM-CSIC, Spain), A. Ferriere (PROMES-CNRS, France) and I. Cañadas (PSA-CIEMAT, Spain) and to use higher powered solar facilities such as the solar furnaces of PSA and the PROMES laboratory.

The aim of our research was not just to make inroads on the use of new non-contaminating technologies, which resolve environmental issues arising from high temperature metallurgy, but also to increase scientific knowledge about the mechanisms involved in these processes carried out at high temperatures under non-equilibrium conditions. In the studies we have conducted to date we have seen a clear activating effect in CSE which results in treatment times that are shorter, and which add to the efficiency of the process as well as increase in the quality of the modified surface. This is due to, among other factors, the properties of solar radiation. The visible solar spectrum extends from the wavelengths between 400 and 700 nm where most metals present greater absorbance, making the processes more energy efficient. In figure 2 (Pitts et al., 1990) the solar spectrum is compared to the absorbance values of the different wavelengths of iron and copper. The figure also includes the wavelength at which certain lasers (those which are habitually used in treating materials) operate. Here we see the high absorbance of iron for the more energetic wavelengths of the solar spectrum, and that its absorbance is low at the wavelengths, which the most common lasers use.

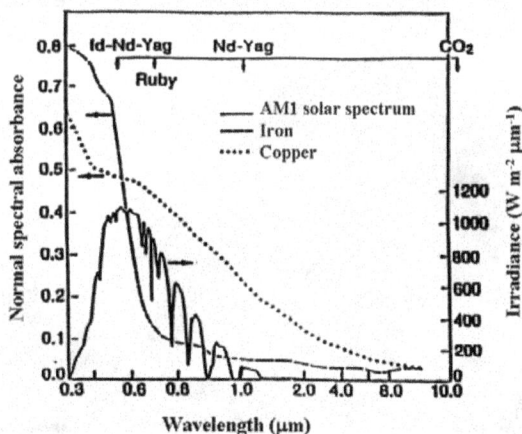

Fig. 2. Solar spectrum (Pitts el al. 1990).

Although the use of solar energy for industrial applications suffers a disadvantage due to its intermittent nature, it should be noted that according to Gineste (Gineste et al. 1999) in Odeillo where the Felix Trombe Solar Furnace Centre is located, the peak value of the direct normal irradiation is 1100 W.m^{-2} and it exceeds 700 W.m^{-2} during 1600 hours per year and 1000 W.m^{-2} during only 200 hours per year. In Ciudad Real, Spain, at latitude 38°, the availability of the solar energy reported by the Spanish "Instituto Nacional de

Meteorologia" (Font Tullot, 1984), is 11% higher than in Odeillo. Direct solar radiation measured with a pyrheliometer between 19 June and 31 August, 2009, at the ETSII-UCLM (Ciudad Real, Spain) registered values higher than 950W.m^{-2} for 20% of the days and higher than 800 W.m^{-2} for 97% of the days. The peak value has been attained in this period was 976 W.cm^{-2} .

2. Experimental Installations

There are various types of installations for concentrating solar energy. One way of classifying these installations uses the concentration process as a reference for differentiating between the different types. In this manner we can distinguish between installations which use reflection and those which use refraction.

Reflection installations

Reflection installations use mirrors to concentrate solar energy producing one diversion (direct concentrators) or several diversions (indirect concentrators) of the radiation. The light is reflected along the entire spectrum of wavelengths, since the mirror does not absorb anything. Direct concentrators are cylindrical parabolic mirrors and dish parabolic reflectors. First one uses the heat energy generated mainly to heat the fluids which circulate through the conduit located in the reflector focal line (Figure 3). Dish parabolic reflector may be full-surface parabolic concentrators when the entire surface forms an approximately parabolic shape or multifaceted concentrators composed of various facets arranged in a parabolic structure that reflects the solar radiation concentrating it in its focal point. The concentration factor depends on the size, aperture and quality of the surface. The solar radiation hitting the focal point has a Gaussian distribution and its energy efficiency is very high due to the high concentration.

Fig. 3. Cylindrical parabolic concentrators at the PSA (Almería Solar Plant).

The indirect concentrators are mainly the solar furnaces. They are systems that take advantage of the thermal energy generated by the sun for use in applications requiring medium to high temperatures. They are indirect concentrators that produce several diversions of the radiation through optical systems specially designed to deflect the incident light. To deflect the radiation, they use mirrored heliostats, completely flat surfaces that deflect the direct solar radiation. They are composed of flat reflective facets and have a sun-tracking system on two axes. Given that a single heliostat is usually totally flat, it does not concentrate. Therefore, a field of heliostats pointed towards a parabolic concentrator is used for this purpose (Fig. 4). The power concentrated may be regulated through an attenuator which adjusts the amount of incident solar light entering.

Fig. 4. Parabolic reflector at the PSA (Almería Solar Plant, Spain)

When the heliostat field is pointed towards a tower (Figure 5) is a direct concentrator because this system produces only one diversion of the solar radiation.

Fig. 5. Heliostat field with a central tower Solar Two, in Barstow, California

Refraction installations

In these installations solar light travels through a concentrator device that redirects the light towards its axis. These types of installations absorb part of the wavelength of the solar light. The most common way of concentrating solar radiation is through the use of converging lenses, which concentrate radiation in its focal point. Conventional lenses would need to be too large and too expensive to make them worthwhile for concentrating solar radiation at the required levels. An alternative to these types of lenses are Fresnel lenses, which serve the same function, but are much lighter and cheaper.

In Fresnel lenses, the curve of the surface is composed of a series of prisms or facets, in such a way that each of them refracts the radiation in the same manner as the surface of which they are a part. This is why a Fresnel lens functions like a conventional lens. The different polymers used in the manufacture of the lens determine the part of the spectrum in which it will be effective, and therefore, its applications. The lenses used for concentrating solar radiation are made of acrylic, rigid vinyl, and polycarbonate. Figure 6 shows how the facets of a Fresnel lens can be created from a conventional lens.

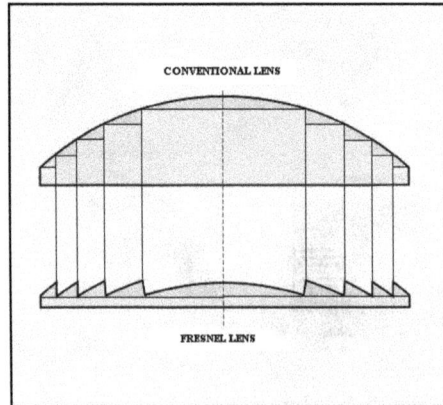

Fig. 6. Diagram of Fresnel lens

There are several research laboratories that use solar installations to experiment and study materials at high temperatures (higher than 1000°C). Table 1 lists the solar installations in operation across the globe, among which is the installation at ETSII in Ciudad Real.

Country	Location	Technology	Maximum power density (kW/m²)	Power (kW)
China	Guangzhou	Parabolic concentrator*	30000***	1.7
France	Odeillo, CNRS	Solar Furnace*	16000	1.5
		Solar Furnace	10000	1000
		Solar Furnace	4700	6
	Odeillo, DGA	Solar Furnace *	6000	45
Germany	Cologne, DLR	Solar Furnace	5200	22
Spain	Almería, PSA-CIEMAT	Solar Tower*	1000-2000	3360-7000
		Solar Furnace *	2500	60
	Madrid, CENIM-CSIC	Fresnel lens*	2640	0.6
	Ciudad Real, UCLM	Fresnel lens*	2640	0.6
Switzerland	Villigen, PSI	Solar Furnace	5000	45
		Solar Furnace	4000	15
		Parabolic concentrator	4000	70
Ukraine	Ac. of Science	Parabolic concentrator *	2500	-
USA	Alburquerque, Sandia	Solar Furnace *	3000	25
	Denver, NREL	Solar Furnace *	2500-20000**	10
	Minneapolis, Univ. Minn.	Solar Furnace	7000	6
Uzbekistan	Tashkent	Solar Furnace	17000	1000

*Used in the surface modification of materials (papers published), **Used as secondary concentrator, ***Calculated values

Table 1. Solar Installations in the World (Rodríguez, 2000).

2.1 Fresnel lens

The installation is on the roof of the Escuela Técnica Superior de Ingenieros Industriales building in the UCLM in Ciudad Real (Figure 7). The lens is affixed in a metal structure, and has a single-axis sun tracking system, connected to a software system in which the different data generated by the experiment can be collected, such as the values of different thermocouples. It also has a pyrheliometer which measures the direct incident solar radiation over the course of the day. The geometry of the lens is circular, with a 900 mm diameter and centre that is 3,17 mm thick. It is made out of acrylic material, which gives it a long useful life with low maintenance. The specification of the lens was determined in previous studies (Ferriere et al., 2004) which allowed the measurement of the concentration factor along the focal axis. The focal point of the lens is 757 mm from its centre. This is the point where the greatest density of energy is reached. The lens concentrates direct solar energy by up to 2644 times (maximum value at the focal point), which means that for exposure of 1000 W/m² the maximum power density at the focal point is 264 W/cm².

Fig. 7. Fresnel lens at the ETSII (Ciudad Real, Spain).

The density of the solar radiation has a Gaussian distribution in function of the distance from the focal point within the focal plane. This variation is what allows us to choose the temperature to be used for the experiment. We can control the energy density of the solar radiation, adjusting the distance of the sample in the Z axis. (Figure 1).

The Fresnel lens has a reaction chamber where experiments can be carried out in a controlled atmosphere. The reaction chamber is features a quartz window and a refrigeration system. In order to measure the temperature a thermocouple is welded to the bottom of the samples.

2.2 Solar Furnace

The second installation used on a regular basis for generating concentrated solar energy is the Solar Furnace of Almería Solar Plant (PSA), which belongs to the Centro de Investigaciones Energéticas, Medioambientales y Tecnológicas (CIEMAT, in English, Centre of Energy, Environmental and Technological Research). The solar furnace consists of a heliostat which tracks the sun and reflects the solar rays onto a parabolic mirror. The furnace of PSA has a heliostat of 160m² composed of 28 flat facets which reflect solar rays perpendicular and parallel to the optic axis of the concentrator and continuously tracks the sun through a tracking system with two axes (Fig. 8). The mirrors have reflectivity of 90%.

Fig. 8. Heliostat of the PSA solar furnace (Almería Solar Plant, Spain).

The concentrator disk is the main component of the solar furnace (Fig. 9). It concentrates the incident light of the heliostat, multiplying the radiant energy in the focal zone. Its optic properties especially affect the distribution of the distribution of the flow on the focal zone. It is composed of 89 spherical facets covering a total surface area of 98,5 m^2 and with a reflectivity of 92%. Its focal distance is 7,45 m. The parabolic surface is achieved with spherically curved facets, distributed along five radii with different curvatures, depending on their distance from the focal point.

Fig. 9. Concentrator disc of PSA

The attenuator (Fig.10) consists of a set of horizontal louvers that rotate on their axes regulating the entry of incident solar light hitting the concentrator. The total energy on the focal zone is proportional to the radiation that passes through the attenuator. The concentration and distribution of the power density hitting the focal point is key factor in a solar furnace. The characteristics of the focus with the aperture 100% opened and solar radiation of 1000 W/m^2 are: peak flux: 3000 kW/m^2, total power: 58 kW, and a focal diameter of 25 mm. In this case, the reaction chamber also allows work to take place in a controlled atmosphere. The chamber also has a quartz window which allows concentrated solar energy to enter and also allows researchers to monitor the experiment using different kinds of cameras (digital and IR).

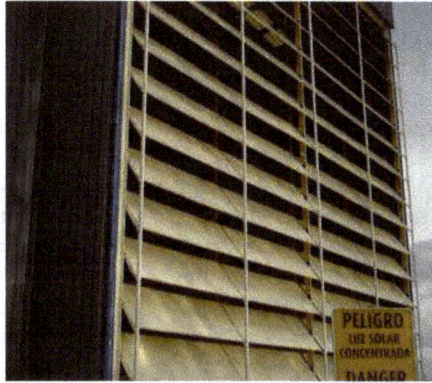

Fig. 10. Attenuator of the solar furnace of PSA (Almería Solar Plant)

2.3 Solar Furnace at PROMES-CNRS
Another solar facility used in our research is the 2kW parabolic solar furnace at the PROMES-CNRS laboratory (France). The furnace is composed of one heliostat and a parabolic reflector with a 2 m diameter. The parabolic concentrator has a vertical axis which allows the samples used in the experiments to rest in a horizontal position without the need to add more optical diversion systems to the device. Furthermore, given that the parabolic reflector is the only mirror which is not faceted, it has a higher quality optical properties and allows a greater concentration factor that that which is obtained with the faceted reflectors. The focal zone behind the second reflection has a diameter of 15 mm and Gaussian distribution with the maximum energy at the centre, of 16.000 times the impinging solar radiation.

3. Surface hardening of steels by martensitic transformation.

Surface quenching is a widely used treatment by the industry to harden and improve wear resistance of the steel pieces. These types of treatments may be carried out using conventional heating methods (flame and electromagnetic induction) and high-density energy beams (laser, electron beams, plasmas, etc.). In all cases the source should be sufficiently powerful to guarantee that only the surface layer of the piece heats up to a higher temperature than the austenizing temperature. After cooling off, only the zones which were previously austenized will have undergone the martensitic transformation that results in the hardening. In the internal zones, where no microstructural transformations would have taken place, the mechanical properties would remain unchanged. Therefore, the end result is pieces that combine a high degree of hardness and toughness and greater resistance to wear.

Of all the different types of modifications and treatment of materials carried out in the solar furnaces, surface hardening of ferrous alloys has been the most widely studied. Since the first study was published by Yu and others in 1982, several research groups have been exploring the possibilities of this process (Maiboroda et al., 1986; Stanley et al., 1990;

Ferriere, 1999). This first study showed how the high concentrations obtained in the solar focal area of a parabolic concentrator with a 1.5 m diameter produced self-quenching in a surface zone 0.5 mm deep and 5 mm in diameter in a steel piece after a second of exposure to solar radiation (Yu et al. 1982). In addition, the initial investigations show how localised treatments can be carried out on industrial pieces with complicated geometries by moving the sample with respect to the focal area of the furnace (Yu et. al, 1983) (Zong et al., 1986).

In Europe, the first experiments were carried out in the 1990's by a group led by Prof. Vázquez of CENIM-CSIC (Spain). The research carried out was highly important because it demonstrated the viability of using the different types of solar facilities available at the Almeria Solar Plant to surface harden steel pieces. The experiments were carried out using the SSPS-CRS facility, which comprises a heliostat field and a central tower (Rodríguez et al., 1995), and the Parabolic Solar Furnace which comprises a group of heliostats and a faceted parabolic concentrator (Rodríguez et al., 1997). The results indicated that under the best conditions of direct solar radiation it was possible to obtain homogeneous quenched layers between 1 and 3 mm thick with heating times of between 30 and 60 seconds. The study was completed using a Fresnel lens with a 900 mm diameter which was available at the Instituto de Energías Renovables of CIEMAT in Madrid (Rodríguez et al., 1994). The study added to knowledge regarding the advantages and limitations of each one of the facilities. With the facility comprising the central tower and the heliostat field, a surface of 10 cm^2 can be quenched, much more than what is possible using other techniques and types of solar facilities. But with the PSA Solar Furnace and the Fresnel lens it is possible to obtain greater energy densities in the focal area (250-300W.cm^{-2}) depending on the incident solar radiation, which allows for self-quenching in steel alloys.

Using the ETSII-UCLM Fresnel lens described above, our group carried out research on surface hardening steels and cast iron, with the ultimate aim being the discovery of industrial applications for this process. The first experiments consisted of surface hardening through martensitic transformation of three types of steel and a nodular cast iron piece.

In all cases the influence on the treatment of the different variables was assessed: heating rate, maximum temperature reached, cooling medium, size of the treated pieces (diameters: 10 and 16 mm, height: 10 and 15 mm). The study entailed determining the microstructural transformations, the profile of the hardness and the depth of the quenching (total and conventional).

Figure 11 shows the results obtained using a sample with a 10 mm diameter and 10 mm high of tool steel AISI 02, where the homogeneous quenching can be seen along the entire diameter of the test sample, as well as the surface hardness values obtained. Figure 12 shows the results obtained after carrying out a surface quenching treatment of a nodular cast iron piece.

In addition, studies were carried out to assess the possibility of carrying out localised treatments on the surfaces of pieces that required greater hardness and resistance to wear. The microhardness curves in Figure 13 show the effect of heating time on the diameter of the quenching zone of a 1 mm plate of martensitic stainless steel AISI 420.

Due to the fact that the heating conditions depend on the direct solar radiation it is necessary to have a predictive tool that can set the treatment conditions in function of the direct solar radiation present. To this end, a finite element model (FEM) (Serna & Rodríguez, 2004) has been developed which gives the distribution of the temperatures of the pieces during treatment. The model takes into account both the Gaussian distribution of the energy density in the focal area and the variation in the temperature of the phase transformation of the steel in function of heating speed.

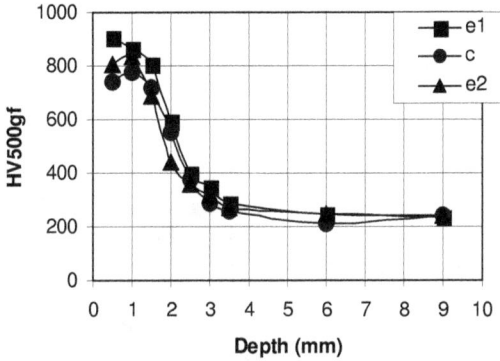

Fig. 11. Surface quenching of 10 mm high test sample after 45 seconds of heating.

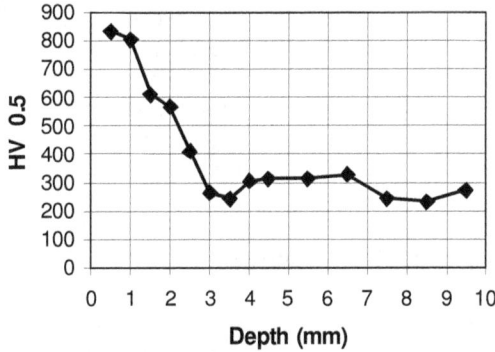

Fig. 12. Microhardness profile of a nodular cast iron piece heated for 40 seconds in the focal area of a Fresnel lens.

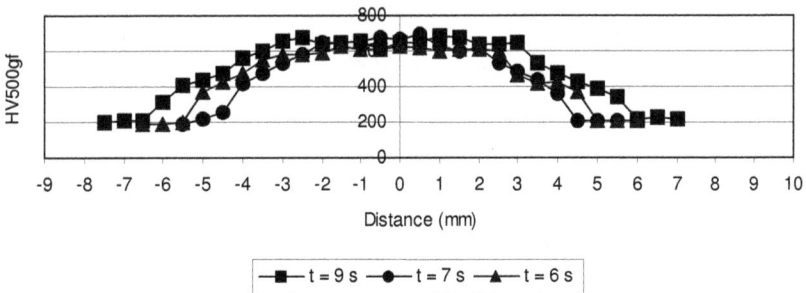

Fig. 13. Influence of heating time on the diameter of the quenching zone of a 1 mm thick plate of martensitic stainless steel.

The size of the focal area of the lens limits the possible application to small pieces (surface areas of between 50 mm^2 and 100 mm^2, depending on the maximum required temperature), or to localised treatments on larger sized pieces. Obviously treatments and modifications of metallic materials with small surface areas are carried out on an industrial scale for a large number of applications, but there are few bibliographic references concerning specific applications for localised treatments using solar facilities.

4. Hardening through surface melting of cast iron

The surface melting treatment of grey cast iron leads to the formation of superficial layers with excellent resistance to wear. The rapid cooling from the melted state gives rise to the formation of extremely hard white cast iron with great resistance to wear. Given this profile, this type of melting process is an excellent candidate for machine pieces that are subject to movement and vibrations and which are in contact with other components. At present, industrial processes are using various heating techniques such as TIG, electromagnetic induction, and electron or laser beams, among others, in order to carry out this type of surface melting treatment.

Recent experiments at the ETSII show how it is possible to use concentrated solar energy to carry out at the melting treatments on industrial cast iron pieces. The study to date has centred on hardening through surface melting and quenching of camshafts used in the automobile industry and manufactured with grey cast iron. In order to carry out the treatment a specimen support device was created which allowed the cam to spin in the focal plane of the Fresnel lens. The results obtained are compared with those of a camshaft manufactured using conventional industrial processes (TIG). Figure 14 shows the hardness profiles of the hardened area of the two pieces, one that was hardened with concentrated solar energy, and the other through TIG (both made of the same cast iron). The graph shows us that the camshaft treated with CSE attained a far greater hardness and had a smaller heat affected zone. In addition, the surface finishing attained from the solar treatment is better than that obtained after treatment with TIG, which would translate into lower costs (Figure 15). Current research is focused on automating the system in such a way that the entire camshaft may be treated continuously.

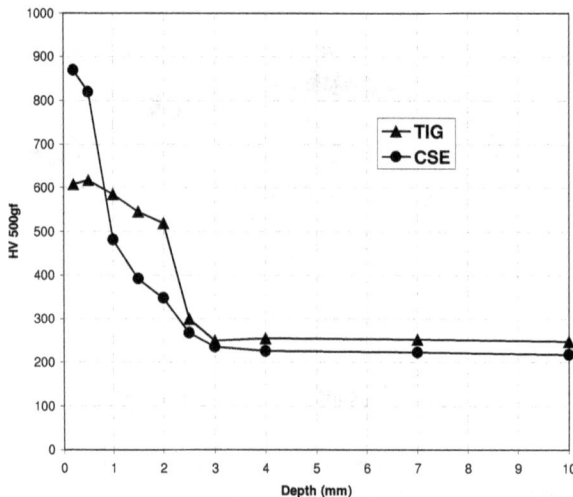

Fig. 14. Hardness profile of samples treated by TIG vs. CSE

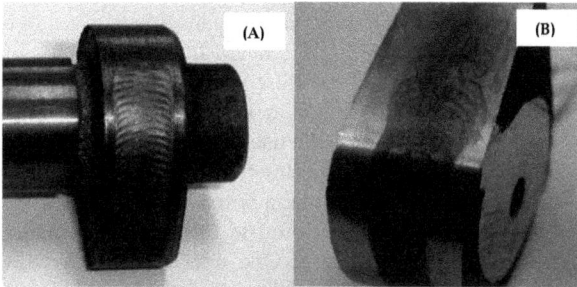

Fig. 15. Surface finishing after TIG (A) or CSE (B) treatment.

5. Cladding of stainless steel and intermetallic compounds onto steel substrates.

There are many types of surface modification techniques that aim to improve the corrosion and oxidation resistance. One widely studied surface modification technique is cladding. A material with the desired properties is melted on the base metal by means of an energy beam. The mixture between the coating material and the base metal must be as small as possible in order to guarantee the original properties of the coating. In this way it is possible to use a cheap structural material and to coat it with another that confers its surface the desired properties.

The literature contains several references that describe the use of different solar installations to obtain cladding coatings. In the USA Pitts et al., (Pitts et al., 1990) obtained cladding coatings on stainless steel. Subsequently, in Spain, Fernandez et al. performed Ni cladding on steel at the Almeria Solar Plant (Fernandez et al., 1998).

We have studied the possibility of obtaining cladding coatings using the parabolic solar furnace of the PROMES-CNRS laboratory previously described (Ferriere et al., 2006). The high energy densities obtained with the solar beam allowed stainless steel and NiAl to be cladding coated through rapid melting-solidification of powders pre-deposited on carbon steel samples. Coatings have been processed in tracks by scanning the concentrated solar beam across the specimen surface with the aim of modifying larger areas than are possible with a stationary treatment. The scanning process is performed by moving the specimen at a controlled speed that depends on the direct solar irradiation. The coatings processed are homogeneous, adherent and have low porosity. In addition, the formation of dendritic microstructures results in increased electrochemical corrosion resistance.

The fundamental disadvantage that has been encountered in this research is the difficulty of achieving a coating with a composition close to that of the initial powder while at the same time guaranteeing good adhesion to the substrate. A possible solution to this problem consists of using a powder injector (nozzle), in order to carry out the process in one single step as is habitual in the case of laser cladding.

6. Salt-bath nitriding of steels.

It is possible to harden the surface of different kind of steels using a novel technology that combines the use of non-contaminant salts with the activator effect of the concentrated solar energy. Groundbreaking research (Shen et al. 2006a), (Shen et al. 2006b) has studied the possibility of the substituting highly contaminating cyanide salts used in liquid nitriding for common salt KNO_3, which avoids this high toxicity. In line with this innovative research vector, our research has tried to explain the different mechanisms through which nitrogen from the salt is introduced into the steel matrix, thereby achieving the desired surface hardening. In addition, we are researching the use of concentrated solar power as an energy source (Herranz & Rodríguez 2008). The experiments were carried out using the Fresnel lens at the ETSII and the solar furnace at the Almeria Solar Plant. The results were compared to those that were obtained using an electric muffle furnace. Steels with a wide range of characteristics were selected for the research. One was a relatively cheap, low alloy steel with low carbon content, the AISI 1042. The other steel used was the high-speed tool steel M2 which is commonly used in industry. These two types of steel were used to highlight the different characteristics made evident during the nitriding process. The treatment resulted in the surface hardening of the two steels through the interstitial diffusion of nitrogen in the steel network and in some samples nitrides were formed. The exhaustive study of these results interprets the nitriding mechanism that occurred in each steel. In addition, the traditional treatment times were reduced to a large degree. This evidences the viability of this nitriding process using concentrated solar energy (CSE).

The maximum surface hardening of AISI 1042, which is not a conventional steel to use for nitriding, increased 61% with respect to its nominal value. Nitrided M2 steel attained a surface hardness of 900HK (Herranz & Rodríguez, 2007) (Fig. 16).

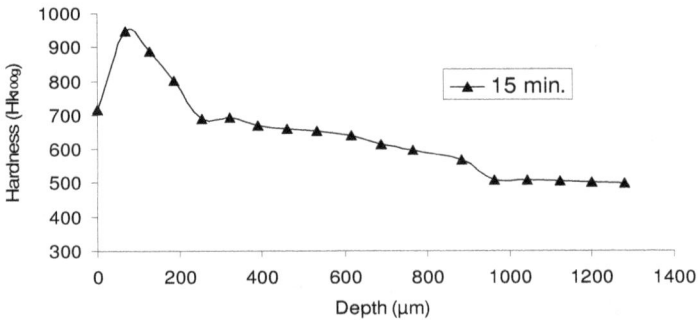

Fig. 16. Hardness profile obtained in a nitrided M2 piece using a treatment of nitrate salts during 15 minutes using concentrated solar energy.

The diffusion layer obtained in both steels submitted to the nitriding process in KNO_3 salts are greater than those obtained in earlier experiments with other nitriding methods. Therefore, the maximum diffusion layer obtained in AISI 1042 steel has an approximate thickness of 1300 µm and 550 µm in the M2 steel.

The other fundamental conclusion of our research is that concentrated solar energy activates the process, notably reducing the treatment times. Treatment time was greatly reduced in the treatment of the AISI 1042 using the Fresnel lens. Only 35 minutes was needed, versus the 90 minutes needed in a muffle furnace for the same treatment. This reduction is especially notable if we compare it with other conventional nitriding methods, such as with plasma, which requires 3 hours. In the case of the nitriding of the M2 sample, the reduction in treatment time using concentrated solar energy is more significant. With a Fresnel lens, only 60 minutes was needed (a muffle furnace requires 3 hours). The reduction was even more remarkable in the case of PSA (Almería Solar Plant), which only needed 15 minutes of treatment.

The nitriding of low alloy steel characterised by its free interstices, such as AISI 1042 steel, occurred primarily through the interstitial diffusion of nitrogen coming from the salts towards the interior part of the steel piece. In the nitriding mechanism for the M2 high-speed steel, high alloy tool steel, part of the nitrogen originating from the salts was introduced through interstitial diffusion towards the interior of the steel, while another part formed iron nitrides and/or iron alloy nitrides. These compounds contribute to the surface hardness.

7. Gas nitriding of titanium alloys.

In recent years, Ti alloys have been widely studied due to their properties: low density, high melting point, good mechanical properties, high corrosion and oxidation resistance and biocompatibility. These properties explain their appeal to the aerospace industry and as biomaterials. However their use in these fields remains limited because of their poor tribological properties. These problems can be overcome by changing the nature of the surface using thermochemical treatments. In the case of the Ti alloys the most widely surface treatment used is the nitriding. There are different ways to nitride the surface. One of the most frequently method is the gas nitriding.

Gas nitriding is a diffusion process that involves heating the surface at high temperatures for a long time. In most cases, the equipment is expensive both start up and to maintenance. For this reason there is significant interest in developing more economic and energy efficient systems. Nitrogen is soluble in titanium and forms a interstitial solid solution, resulting in a hardening by solid solution caused by deformation in the crystalline network. It is important to assure that no oxygen is present in the process in order to prevent the formation of TiO_2 oxides on the piece. The types of dissolution of nitrogen in the titanium may be seen in the equilibrium diagram of the Ti-N system (Fig. 17) (Wriedt &

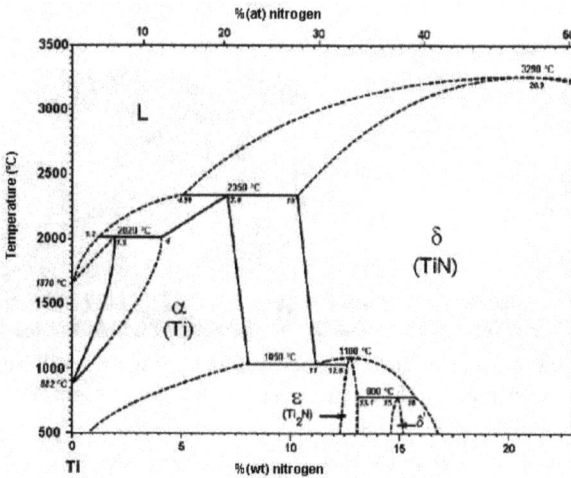

Fig. 17. Phase diagram Ti-N. (Wriedt & Murray, 1987).

Murray, 1987). We observe a solid phase α solution with a high degree of solubility of nitrogen (from 0 to 8% in weight) which remains stabilised at high temperatures. At levels of over 8% of nitrogen (in weight) the alpha phase saturates and intermetallic compounds are obtained, such as Ti_2N and TiN. These nitrides have very high hardness values and give the maximum hardness to the surface layer.

Existing studies indicate that there are some problems related to the use of this treatment in the traditional manner, which uses electric furnaces, such as the need for long treatment times of over 16 hours and at high temperatures of over 1000° C. In contrast, the use of concentrated solar energy obtains high temperatures at high heating and cooling rates (that allow the creation of non-equilibrium microstructures) with short treatment times. We have found that due to the photoactivation capability of the concentrated solar energy, the duration of the process can be greatly reduced. To our best knowledge, the study by Vazquez in 1999 is the only significant research that has been conducted in this area (Sánchez Olías et al, 1999).

Our research focuses on the gas nitriding of the Ti6Al4V alloy. The base material has a biphasic microstructure (α+β) and a hardness of 400 HK. Tests involved applying heat of between 1000° C and 1200°C for between 5 to 30 min. In order to measure the temperature, a thermocouple was welded to the bottom of the samples and the samples were situated in the reaction chamber where the nitrogen atmosphere was controlled (Fig. 18). The chamber has a quartz window which permits the entrance of the concentrated solar energy and also it allows to observe and record the experiment by different kind of cameras (digital and IR).

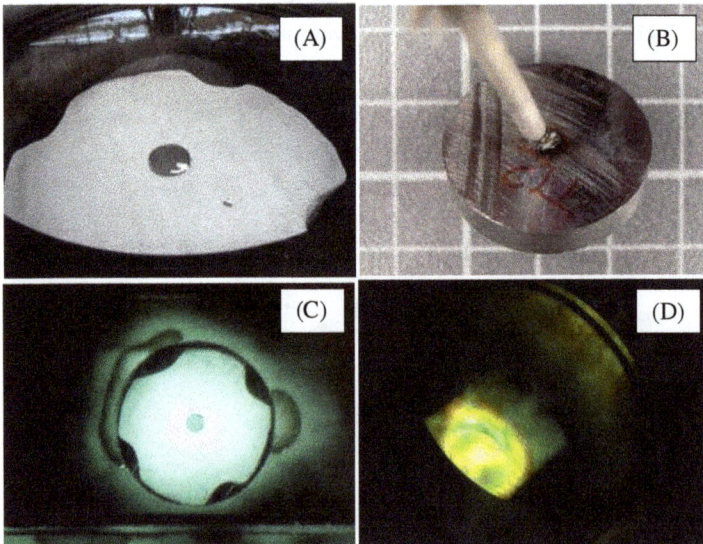

Fig. 18. (A) The samples in the reaction chamber. (B) Sample with the thermocouple. (C) Digital record of the experiment in the solar furnace. (D) Direct observation of the experiment in the Fresnel lens.

The results obtained indicate that with the solar facilities significant hardening can be obtained from processes carried out at 1050°C for only 5 minutes. In the microstructure we observed a layer of a different kind of nitrides identified by X-Ray diffraction that increase the hardness of the samples. In addition, nitride layer growth occurred where the treatment time was increased. Besides the hardening due to the formation of nitrides, we also observed a deeper hardening due to the interstitial solid solution of the nitrogen in titanium matrix. As shown in Figure 19 we have been able to distinguish between the different hardened layers. The first is a compound layer, then a diffusion layer that can achieve 400 μm for the longest treatment times and then, the base material.

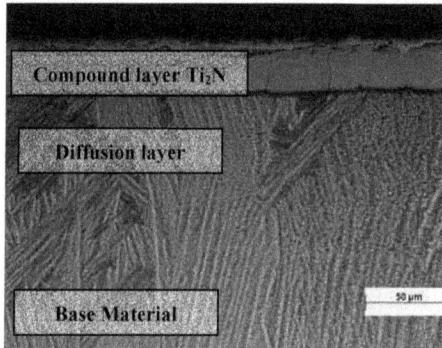

Fig. 19. Microstructure of sample nitrided during 10 min at 1050°C using concentrated solar energy.

The compound layer was identified through X-ray diffraction as Ti_2N, which was the maximum hardness detected (1200HK). The presence of TiN was also detected, but its density was so scant that it was impossible to measure its hardness. In addition, it was evident that the process did not result in a completely continuous layer at this temperature. Compound layers grew considerably as the treatment temperature rose or time was increased. These layers covered the pieces more homogeneously and the diffusion layer was thicker. As shown in figure 20, at 1200°C, after 15 minutes of treatment, two totally continuous layers were formed. According to the X-ray diffraction, the exterior layer was TiN, and the second layer was Ti_2N. Under these parameters, the TiN layer had a maximum hardness of 2600 HK.

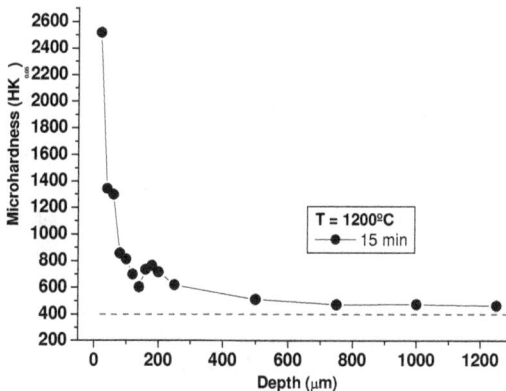

Fig. 20. Microhardness evolution in the samples nitrided during 15 min at 1200°C using CSE

After these experiments, wear resistance was evaluated using a pin on a disc test machine. The tests were made in dry conditions against a ball of aluminium oxide. In the case of the samples nitrided in the Fresnel lens, the samples showed a lower coefficient than the as-received material, with a friction coefficient of 0.5, showing values of 0.2 for treatment of only 15 minutes at 1200°C. In regard to the wear rate, compared with the base material it decreased by two orders of magnitude after only 15 minutes of treatment. The track width was decreased from 1.890 μm in the Ti6Al4V alloy to 180 μm in the nitrided sample. These experiments show the significant potential of this new modification process consisting of gas nitriding with concentrated solar energy. The great reduction of nitriding time can be explained by the photo-activation effect of the concentrated solar energy.

8. Solar sintering of metallic powders.

The feasibility of using a solar furnace for the sintering-consolidation of green parts previously obtained by compaction has been tested. The use of solar furnaces allows materials to be processed at much higher heating and/or cooling rates, which results in better mechanical properties and a quite significant reduction in the total cycle time. The demand for faster and less expensive techniques for processing pieces has resulted in the development of high energy techniques for manufacturing prototypes and carrying out short series tooling, such as laser sintering (Asgharzadeh & Simchi 2005) and capacitor discharge sintering (CDS) (Fais & Maizza 2008). The use of these faster technologies has usually resulted in finer microstructures, while improving mechanical properties and the sintering window. In the search for new sintering systems that overcome this issue, the use of concentrated solar energy (CSE) appears to be an interesting candidate. The CSE is clean, renewable and pollutant free. In spite of its apparent limitations for industrial applications, due to the unpredictable availability of solar radiation, CSE shows a clear activator effect in different processes (Herranz & Rodriguez 2007); (Herranz & Rodríguez 2008).
Earlier research focused on studying the technique's viability for manufacturing WC-10%Co ceramic components and complex ceramics (cordierite) (Guerra Rosa et al., 2002); (Almeida Costa Oliveira et al., 2005); (Cruz Fernandes et al., 2000) using concentrated solar energy. These studies have proven that it is possible to use CSE to obtain pieces with very similar characteristics to those obtained using conventional electric furnaces. In the same line, a preliminary study of a copper system has also been published (Cañadas et al., 2005). The results of these studies all share two characteristics: the good mechanical properties obtained, and the major reduction in total treatment time required.
We have assessed the feasibility of concentrated solar energy for sintering-consolidation of green parts previously obtained using compacted metallic powder. The experiments were carried out using the two facilities described in this chapter; a parabolic solar furnace of the Almeria Solar Plant (PSA) and a Fresnel lens at the Castilla-La Mancha University (ETSII-UCLM). The research focused on the microstructural evolution of the samples treated in both facilities. In this work we compared the results of the sintering process in a N_2-H_2 atmosphere using concentrated solar energy with the conventional furnace results. The main materials we worked with were M2 tool steel, copper-based alloys (such as bronze), and carbon-based steels with alloys such as Astalloy. The energy density obtained in the focal area of the solar furnace or in the focal area of the Fresnel lens is high enough to raise the sample's temperature to sintering levels. To process the samples at different sintering temperatures, experiments were conducted changing the focal length. The samples were

then heated up to the maximum temperature for 30 minutes. The total duration of the sintering process (including the heating and the cooling) was less than 70 minutes in all cases. In order to compare results, the sintering experiments were carried out in a conventional tubular furnace in the same atmosphere, applying conventional sintering temperatures for 30 min (dwell time). (Figure 21).

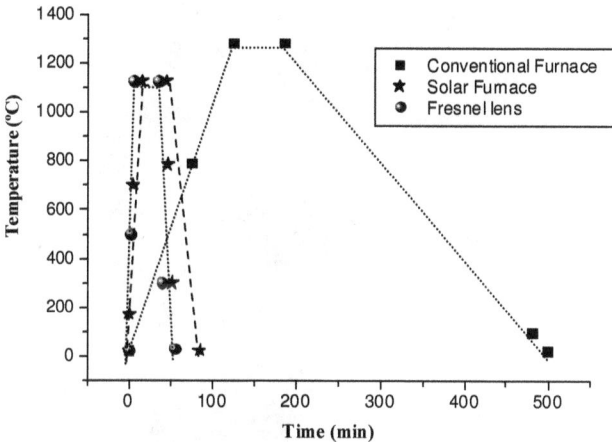

Fig. 21. Cycle in the conventional furnace and in the solar installations comparing the duration of the sintering processes.

We made several interesting findings. For example, we found that the high-speed M2 steel showed major differences in the microstructures obtained, depending on the sintering process used. The microstructure of M2 sintered in a conventional furnace in N_2-H_2 atmosphere at the optimum sintering temperature, around 1290°C, presented a ferrite matrix with some retained austenite, Figure 22. We also observed a significant amount of homogeneously distributed bright rounded M6C carbides (identified as rich in W and Mo). The sintering atmosphere has an important influence on the development of the microstructure because the nitrogen content of the steel slightly increases the optimum sintering temperature. The increase in nitrogen content has no effect on the M_6C carbides but, as pointed out in a previous study by Jauregi (Jauregi et al 1992), the MC carbides (rich in V) transform into MX carbonitrides that appear with grey contrast inside the grains and at the grain boundaries. In the conventional furnace an increase of the temperature up to 1300°C produces an over-sintered microstructure in which a continuous M_6C carbides film around the grain boundaries is observed. The EDX analysis revealed that at high temperatures the MX carbonitrides were able to transform themselves into black-contrast square VN nitrides detected at the grain boundaries. The maximum hardness achieved was 550 HV.

In the case of the samples treated in the PSA solar furnace and with the Fresnel lens, we observed a completed densification. The most remarkable discovery was that for the processes carried out using concentrated solar energy, the process activated at lower temperatures. Figure 23 shows the densification process and the microhardness evaluation. The feasibility of the sintering process using CSE was demonstrated as the parts displayed well-defined necks and greatly reduced porosity.

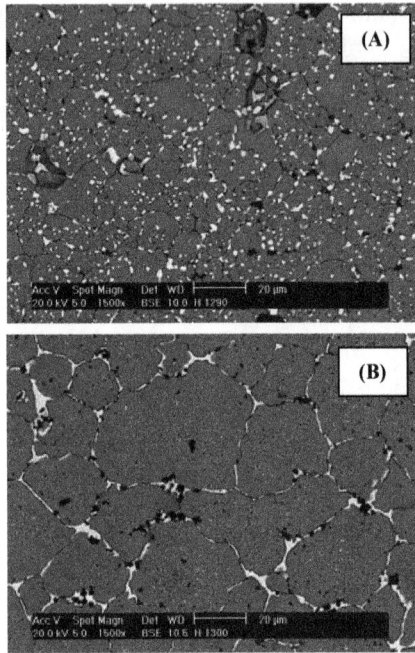

Fig. 22. SEM micrograph of M2 sample sintered at (A) 1290°C and (B)1300°C in nitrogen-hydrogen atmosphere in the conventional furnace.

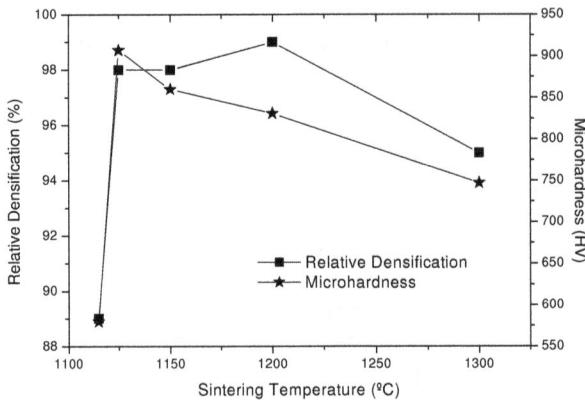

Fig. 23. Densification process of the samples treated in the solar furnace of PSA.

As can be seen in the Fig. 24 (A), at only 1115 °C, some rounded and isolated porosity was observed and a homogeneous distribution of carbides indicates that the sintering process really took place. Microhardness measurements indicate that, at this temperature, the distribution of the carbides was homogeneous and the porosity had disappeared in most of the sample. The microhardness achieved values similar to those of the samples sintered in the conventional furnace (580 HV). The microstructure obtained at this temperature was the same as that obtained at 1260°C in the electric furnace. When the sintering temperature was increased to 1125 °C, Fig. 24 (B), full density was achieved and there was a sharp increase in grain growth. In addition, massive bright carbide segregation was observed, as well as formations corresponding to the eutectic phase, but only in some areas. Most of the grain boundaries did not show carbides, indicating that they underwent a partial dissolution during the sintering process. Based on EDX analysis, we ascertained that the increase of nitrogen in the matrix due to the sintering gas allowed the formation of VN nitrides both inside the particles and along the grain boundaries. In spite of the grain growth, this sample's hardness was over 900 HV.

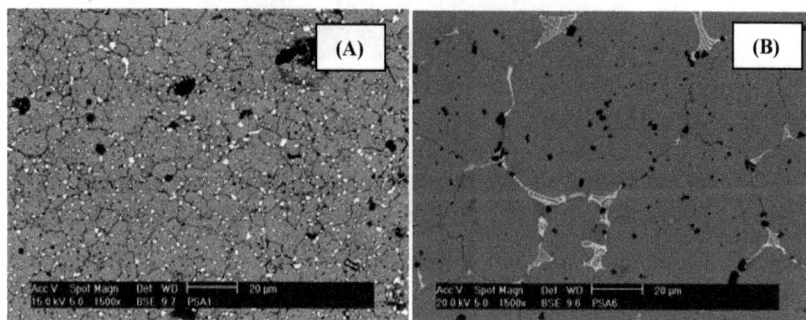

Fig. 24. SEM micrograph of M2 sample sintered at (A) 1115°C and (B)1125°C in nitrogen-hydrogen atmosphere in the solar furnace.

As part of our efforts to explain these results, we carried out several preliminary studies using transmission electron microscopy (TEM). Initial analysis corroborated that the high velocities obtained had caused the formation of vanadium-rich nanometric particles on the order of 400 nanometres, which in certain cases may be complex and contain small 30 nanometre particles rich in other steel alloy elements (W, Mo and V) in their interior. The presence of these particles would explain the high hardness values obtained. However, it would be necessary to carry out a more in depth analysis to fully understand the mechanism responsible for the formation of these small particles.

The results obtained using the Fresnel lens are similar. In figure 25, the large darker particles (identified as VN) stand out in particular and are precipitated at the grain boundaries or close to them, with an average size of 2 μm. In addition, there are a large number of small nanometric particles dispersed throughout the matrix, which once again explain the high hardness values obtained, despite the rapid growth observed in the grain. Lighter contrast carbides (rich in W and Mo) are concentrated at the grain boundaries while larger particles are distributed homogenously throughout the sample. In addition, eutectic M_6C "fishbone" structures can also be observed at some of the grain boundaries.

We would highlight that certain pieces were totally densified at lower temperatures, 150°C lower than with a conventional furnace. Almost full density was attained after 75 minutes in the PSA and after 50 minutes in the Fresnel lens installation, with higher hardness than the conventional microstructures (760-900 HV). The higher heating and cooling rates could explain these results, since the equilibrium phase diagram changes with the heating rate and in these new heating conditions new sintering mechanisms have occurred. The microstructure and the hardness measurements were similar to those found in the M2 system treated by selective laser sintering (Niu & Chang, 2000).

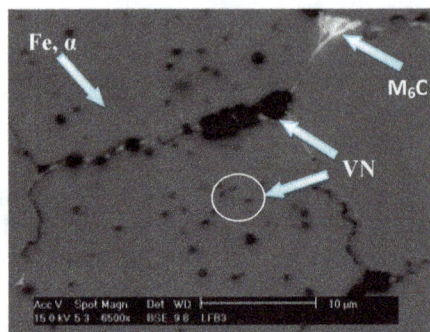

Fig. 25. SEM micrograph of M2 sample sintered in nitrogen-hydrogen atmosphere in the Fresnel lens.

9. Processing of intermetallic coatings through a self-propagating high temperature synthesis process initiated with solar energy (SHS-CSE).

The intermetallic compound NiAl shows attractive properties such as low density, high melting temperatures, high thermal conductivity and good mechanical behaviour at high temperature. Although its low ductility at room temperature could limit its applications as structural material its good tribological properties and excellent oxidation resistance justify its use as a protective coating for metallic components. Cladding, and self-propagating high temperature synthesis (SHS) processes are among the other techniques used currently for coating this intermetallic compound (Matsuura et al., 2000).

SHS is an energy efficient method to process advanced ceramic materials and intermetallic compounds. Discovered by Merzhanov in 1967 (Merzhanov, 1967) it uses the highly exothermic properties of a chemical reaction that are sustained and propagated through a mix of reactants (usually in powder form) in the form of a combustion wave. The energy savings and the economic benefits obtained from SHS are widely acknowledged (see, for example, Deevi & Sikka, 1997).

Concentrated solar energy was first used for the coating process using an SHS reaction in a study by G.P. Rodríguez et al. (Rodríguez et al., 1999) in which they were able to process coatings of NiAl on circular samples with a diameter of 30 mm, using a Fresnel lens. The solar energy triggered a exothermic reaction between the Ni and the Al at ignition temperature (the melting point of Al). The heat released by this reaction was transferred through conduction to the adjacent zones, initiating the reaction in these zones again, for which reason the reaction is considered self-propagated. The heat released (combined with that of the solar beam) melts the compound obtained, resulting in a dense non-sinterized material. This process was optimised by Sierra and others (Sierra & Vázquez, 2005), where the adherence of the coating to the substrate was improved through the electrodeposition of a nickel layer prior to the treatment. The study carried out with the Fresnel lens of the ETSII-UCLM used elemental Ni and Al powders and resulted in adherent NiAl coatings over cast iron. In order to increase the adherence and decrease the porosity that frequently occurs in coating processes that use SHS, a preheating system was designed for the substrate, which also uses solar power.

In addition, the 2 kW solar furnace at the PROMES-CNRS laboratory was used (under the collaboration framework between CENIM-CSIC, PROMES-CNRS and ETSII-UCLM) to obtain NiAl coatings in track forms, processing larger surface areas than those that were processed using the Fresnel lens (Sánchez Bautista et al., 2006). The scanning method is similar to that used when coating surfaces using solar cladding, although the beam-sample interaction times are lower (due to faster movement of the sample). The main difficulty arising when coating surfaces using a solar assisted SHS process lies in controlling the process variables so that the resultant coating is of low porosity, high adherence and with the required chemical composition. A possible solution may involve preheating the substrate. Another challenge is to optimise the processing of coatings over large areas.

10. Conclusions

Concentrated solar energy (CSE) represents an alternative to other types of energy beams for treating and modifying the surfaces of metallic materials. The research conducted by the Metallic Materials group of ETSII-UCLM (Spain) is interesting for two reasons. First, it is breaking new ground in the use of new non-contaminating and environmentally acceptable technologies for processes involving the surface modification of metallic materials at high temperatures. Second, it is increasing scientific knowledge about the mechanisms involved when processing materials at high temperatures under non-equilibrium conditions. In the studies carried out to date, it has been observed that CSE has a clear activator effect, which results both in shorter treatment times and therefore, in increased processing efficiency, and in improved quality of the modified surface.

For high-temperature processes, concentrated solar energy has shown to be highly energy efficient and also competitive in terms of cost. It is especially suitable for countries such as Spain, which has a high number of sunny days per year. Going forward, one of the main challenges for the scientific community will be to develop industrial applications for solar technology, especially in the current context, where energy-efficiency and environmental preservation have become top social priorities.

11. References

Almeida Costa Oliveira, F.; Shohoji, N. ; Cruz Fernandes, J. & Guerra Rosa, L. (2005). Solar sintering of cordierite-based ceramics at low temperaturas, *Solar Energy*, 78, 3, 351-361 ISSN 0038-092X

Asgharzadeh, H. & Simchi A. (2005). Effect of sintering atmosphere and carbon content on the densification and microstructure of laser-sintered M2 high-speed steel powder, *Materials Science and Engineering A*, 403, 290-298, ISSN 0921-5093

Cañadas, I.; Martínez, D.; Rodríguez, J. & Gallardo, J.M. (2005). Viabilidad del uso de la radiación solar concentrada al proceso de sinterización de cobre, *Rev. Metal. Madrid*, Vol. Extr. 165-169, ISSN 0034-8570

Chaudron, G.& Trombe, F. (1973). *Les hautes températures et leurs utilisations en physique et en chimie*, Masson et Cie, ISBN 2225360698; Paris, France.

Cruz Fernandes, J.; Amaral, P.M.; Guerra Rosa, L. & Shohoji, N. (2000). Weibull statistical analysis of flexure breaking performance for alumina ceramic disks sintered by solar radiation heating, *Ceram. Int.* 26, 203–206 ISSN 0272-8842

Deevi, S.C. & Sikka, V.K. (1997) Exo-Melt_ process for melting and casting intermetallics, *Intermetallics*; 5, 17-27, ISSN 0966-9795

Fais, A. & Maizza, G. (2008). Densification of AISI M2 high speed steel by means of capacitor discharge sintering (CDS), *Journal of Materials Processing Technology*, 202, 70-75, ISSN: 0924-0136

Fernández, B.J. ; López, V.; Vázquez, A.J. & Martínez, D. (1998). Cladding of Ni superalloy powders on AISI 4140 steel with concentrated solar energy, *Solar Energy Materials & Solar Cells*, 53, 1-2, 153-161, ISSN 0927-0248.

Ferriere, A.; Faillat, C.; Galasso, S.; Barallier, L.& Masse, J.E. (1999). Surface hardening of steel using highly concentrated solar energy process, *Journal of Solar Energy Engineering*, 21, 36-39, ISSN 0199-6231

Ferriere, A.; Rodríguez, G.P. & Sobrino, J.A. (2004). Flux distribution delivered by a fresnel lens used for concentrating solar energy, *Journal of Solar Energy Engineering*, 126, 1, 654-660, ISSN 0199-6231

Ferriere, A. ; Sanchez Bautista, C.; Rodriguez, G.P. & Vazquez, A.J. (2006). Corrosion resistance of stainless steel coatings elaborated by solar cladding process, *Solar Energy* 80, 10, 1338-1343, ISSN 0038-092X

Flamant, G.; Ferriere, A.; Laplaze, D. & Monty, C. (1999). Solar processing materials: opportunities and new frontiers. *Solar energy*, 66, 2, 117-132, ISSN 0038-092X

Font Tullot, I. (1984). *Atlas de la radiación solar en España*, Instituto Nacional de Meteorología, ISBN 8450505011, Madrid, Spain.

Guerra Rosa, L.; Amaral, P.M.; Anjinho, C.; Cruz Fernandes J. & Shohoji, N. (2002) Fracture toughness of solar-sintered WC with Co additive, *Ceram. Int.* 28, 345–348. ISSN 0272-8842

Gineste, J.M.; Flamant, G.& Olalde, G. (1999). Incident solar radiation data at Odeillo solar furnaces., *Journal de Physique IV*, 9, 623-628, ISSN 1155-4339.

Herranz, G. & Rodriguez, G. (2007). Surface Nitriding of M2 HSS using Nitrate Salt Bath, *Proceedings of Euro PM2007*, Vol. 3, 425-428, ISBN 978-1-899072-30-9, EURO PM 2007, International Powder Metallurgy Conference & Exhibition, Toulouse, Oct. 2007, Ed. The European Powder Metallurgy Association (EPMA)

Herranz, G.& Rodríguez, G. (2008). Solar power drives improved wear resistance in HSS, *Metal Powder Report* 63, 4, (April) 28-29, 31, ISSN 0026-0657

Hunt, L.B. (1982). The First Real Melting of Platinum: Lavoisier's ultimate success with oxygen, *Platinum Metals Rev.*, 26, 2, 79-86, ISSN 0032-1400

Jauregi, S.; Fernández, F.; Palma, R.H.; Martínez, V. & Urcola, J.J. (1992). Influence of atmosphere on sintering of T15 and M2 steel powders, *Metallurgical and Materials Transactions A*, 23, 2, 389-400, ISSN 1073-5623

Maiboroda, V.P.; Pasichniy, V.V.; Palaguta, N.G.; Stegnii, A.I.& Krivenko, V.G. (1986). Special features of local heat treatment of steel 34KhN3MFA in the focal spot of a solar furnace, *Metalloved. i Term. Obrab. Met.*, 1, 59-60, ISSN 0026-0819.

Matsuura, K. ; Jinmon, H. & Kudoh, M. (2000). Fabrication of NiAl/Steel Cladding by Reactive Casting, *ISIJ International*, 40, 2, 167–171, ISSN 0915-1559

Merzhanov, AG. (1967). Thermal explosion and ignition as a method for formal kinetic studies of exothermic reactions in the condensed phase, *Combustion and Flame*, 11, 3, 201-211, ISSN 0010-2180

Niu, H. J. & Chang, I.T.H. (2000). Selective laser sintering of gas atomized M2 high speed steel powder, *Journal of Materials Science*, 35, 1, 31-38, ISSN: 0022-2461

Pitts, J.R.; Stanley, J.T.& Fields, C.L. (1990). Solar Induced Surface Transformation of Materials, In: *Solar Thermal Technology-Research-Development and Applications*, B.P. Gupta & W.H. Trangott, (Eds.), 459-470, Hemisphere Publishing Corporation, ISBN 1560320958, New-York, USA

Rodríguez, G.P.; Vázquez, A.J.& Damborenea, J.J. (1994). Steel heat treatment with Fresnel lenses, *Materials Science Forum*, 163, 133-138, ISSN: 0255-5476

Rodríguez, G.P.; López, V.; Damborenea, J.J. & Vázquez, A.J. (1995). Surface transformation hardening on steels treated with solar energy in central tower and heliostat field, *Solar Energy Materials and Solar Cells*, 37, 1-12, ISSN 0927-0248

Rodríguez, G.P. ; Damborenea, J.J. & Vázquez, A.J. (1997). Surface Hardening of Steel in a Solar Furnace, *Surface and Coatings Technology* 92, 165-170, ISSN 0257-8972

Rodríguez, G.P.; García, I. & Vázquez, A.J. (1999). Materials and coating processing by self-propagating high-temperature synthesis (SHS) using a Fresnel lens, *Journal de Physique IV*, 9; 411-416, ISSN 1155-4339

Rodríguez, G.P. (2000) Modificación de superficies con energía solar. *Ciencia e ingeniería de superficie de los materiales metálicos.* Ediciones CSIC, Colección Textos Universitarios, No 31, 211-226, Madrid. ISBN: 84-00-07920-5.

Sánchez Bautista, C.; Ferriere, A.; Rodríguez, G.P; López-Almodovar, M.; Barba, A.; Sierra, C. & Vázquez, A.J. (2006). NiAl intermetallic coatings elaborated by a solar assisted SHS process, *Intermetallics*, 14 ,10-11, 1270-1275, ISSN 0966-9795

Sánchez Olías, J.; García, I.& Vázquez, A.J. (1999). Synthesis of TiN with solar energy concentrated by a Fresnel lens, *Materials Letters*, 38, 379-385, ISSN 0167-577X

Serna Moreno, M.C. & Rodríguez Donoso, G.P. (2004). Predicciones numéricas del endurecimiento superficial de aceros mediante energía solar concentrada, In: *Métodos Computacionais em Engenharia*, Soares, Batista, Bugida, Casteleiro, Goicolea, Martins, Pina y Rodríguez, (Eds.) 33, APMTAC y SEMNI, ISBN 972-49-2008-9, Lisboa, Portugal.

Shen, Y. Z.; Oh, K. H. & Lee, D. N. (2006a). Nitriding of Interstitial Free Steel in Potassium-Nitrate Salt Bath, *ISIJ International*, 46, 1, 111–120, ISSN 0915-1559

Shen, Y.Z., Oh, K.H. & Lee, D.N. (2006b). Nitrogen strengthening of interstitial-free steel by nitriding in potassium nitrate salt bath, *Materials Science and Engineering A* 434, 314–318, ISSN 0921-5093

Sierra, C. & Vázquez, A.J. (2005). NiAl coatings on carbon steel by self-propagating high temperature synthesis assisted with concentrated solar energy: mass influence on adherence and porosity, *Solar Energy Materials & Solar Cells*, 86, 33-42, ISSN 0927-0248

Stanley, J.T.; Fields, C.I.& Pitts, J.R. (1990). Surface treating with sunbeams, *Adv. Mat. Proc.*, 12, 16-21; ISSN 0882-7958

Vázquez, A.J. & Damborenea, J.J. (1990). Aplicaciones de la energía solar al tratamiento de materiales metálicos. Resultados preliminares. *Rev. Metal. Madrid*, 26, 3, 157-163, ISSN 0034-8570

Wriedt, H.A. & Murray, J.L. (1987). The N-Ti system, *Bulletin of Alloy Phase Diagrams*, 8, 4, 378, ISSN 0197-0216

Yu, Z.K.; Zong, Q.Y.& Tam, Z.T. (1982). A preliminary investigation of surface hardening of steel and iron by solar energy, *Journal of Heat Treating*, 2, 4, 344-350, ISSN 0190-9177

Yu, Z.K.; Zong, Q.Y.& Tam, Z.T. (1983). A further investigation of surface hardening of iron and steel by solar energy, *Journal of Heat Treating*, 3, 2, 120-125, ISSN 0190-9177

Permissions

All chapters in this book were first published in SE, by InTech Open; hereby published with permission under the Creative Commons Attribution License or equivalent. Every chapter published in this book has been scrutinized by our experts. Their significance has been extensively debated. The topics covered herein carry significant findings which will fuel the growth of the discipline. They may even be implemented as practical applications or may be referred to as a beginning point for another development.

The contributors of this book come from diverse backgrounds, making this book a truly international effort. This book will bring forth new frontiers with its revolutionizing research information and detailed analysis of the nascent developments around the world.

We would like to thank all the contributing authors for lending their expertise to make the book truly unique. They have played a crucial role in the development of this book. Without their invaluable contributions this book wouldn't have been possible. They have made vital efforts to compile up to date information on the varied aspects of this subject to make this book a valuable addition to the collection of many professionals and students.

This book was conceptualized with the vision of imparting up-to-date information and advanced data in this field. To ensure the same, a matchless editorial board was set up. Every individual on the board went through rigorous rounds of assessment to prove their worth. After which they invested a large part of their time researching and compiling the most relevant data for our readers.

The editorial board has been involved in producing this book since its inception. They have spent rigorous hours researching and exploring the diverse topics which have resulted in the successful publishing of this book. They have passed on their knowledge of decades through this book. To expedite this challenging task, the publisher supported the team at every step. A small team of assistant editors was also appointed to further simplify the editing procedure and attain best results for the readers.

Apart from the editorial board, the designing team has also invested a significant amount of their time in understanding the subject and creating the most relevant covers. They scrutinized every image to scout for the most suitable representation of the subject and create an appropriate cover for the book.

The publishing team has been an ardent support to the editorial, designing and production team. Their endless efforts to recruit the best for this project, has resulted in the accomplishment of this book. They are a veteran in the field of academics and their pool of knowledge is as vast as their experience in printing. Their expertise and guidance has proved useful at every step. Their uncompromising quality standards have made this book an exceptional effort. Their encouragement from time to time has been an inspiration for everyone.

The publisher and the editorial board hope that this book will prove to be a valuable piece of knowledge for researchers, students, practitioners and scholars across the globe.

List of Contributors

Christos D. Papageorgiou
National Technical University of Athens, Greece

Valentina A. Salomoni and Carmelo E. Majorana
University of Padua, Italy

Giuseppe M. Giannuzzi, Adio Miliozzi and Daniele Nicolini
ENEA – Agency for New Technologies, Energy and Environment, Italy

Feng Zhu, Jian Hu, Ilvydas Matulionis and Arun Madan
MVSystems, Inc., 500 Corporate Circle, Suite L, Golden, CO, 80401, USA

Todd Deutsch
Hawaii Natural Energy Institute (HNEI), University of Hawaii at Manoa, Honolulu, HI 96822, USA

Nicolas Gaillard and Eric Miller
National Renewable Energy Laboratory (NREL), Golden, CO 80401, USA

J. C. Bernède, L. Cattin and M. Morsli
Université de Nantes, Nantes Atlantique Universités, LAMP, EA 3825, Faculté des Sciences et des Techniques, 2 rue de la Houssinière, BP 92208, Nantes, F-44000, France

A. Godoy
Facultad Ciencias de la Salud, Universidad Diego Portales. Ejército 141. Santiago de Chile, Chile

F. R. Diaz and M. A. del Valle
Facultatd de Quimica, PUCC, Casilla 306, Correo 22, Santiago, Chile

Oleksandr Malik and F. Javier De la Hidalga-W
National Institute for Astrophysics, Optics and Electronics (INAOE), Mexico

Gytis Juška and Kęstutis Arlauskas
Vilnius University, Lithuania

Gemma Herranz and Gloria P. Rodríguez
University of Castilla La Mancha, ETSII
Metallic Materials Group
Avda. Camilo José Cela s/n. 13071, Ciudad Real, Spain

Index